$x + y$

EUGENIA CHENG is Scientist in Residence at the School of the Art Institute of Chicago and Honorary Visiting Fellow at City, University of London. A concert pianist, she also speaks French and Cantonese, and is the author of *The Art of Logic*, *How to Bake Pi* and *Beyond Infinity*, the last of which was shortlisted for the 2017 Royal Society Science Book Prize.

Also by Eugenia Cheng

The Art of Logic

Beyond Infinity

How to Bake Pi

$x + y$

A Mathematician's Manifesto
for Rethinking Gender

EUGENIA CHENG

P

PROFILE BOOKS

This paperback edition published in 2021

First published in Great Britain in 2020 by
Profile Books Ltd
29 Cloth Fair
London
EC1A 7JQ

www.profilebooks.com

1 3 5 7 9 10 8 6 4 2

Typeset in Sabon by MacGuru Ltd
Printed and bound in Great Britain by
CPI Group (UK) Ltd, Croydon CR0 4YY

A CIP catalogue record for this book is available from the British Library.

ISBN 978 1 78816 041 4
eISBN 978 1 78283 443 4

To and for
the ever-changing congressive constant
Gregory Peebles

Contents

Preface

As I write this, the world is in a state of great uncertainty. It is the middle of March 2020, and a global pandemic is in the process of closing down society as we know it.

It seems impossible to begin this book without mentioning that, but also impossible to know what to say that will still be relevant when the book reaches your hands. The situation is developing faster than we can keep up with, and we don't know what is coming in the approaching days, weeks, months. The range of possibilities is unthinkably wide.

However, one thing that this worldwide crisis has grimly illuminated is a sharp opposition between those who think as individuals and those who think as a community. We see individuals deciding how much risk of infection they are going to take, as if this decision is their own private one, affecting them alone. What we need is individuals acting on behalf of the community, to reduce the risk to the community and protect our collective health.

This contrast between individualistic thinking and community-minded thinking is the major theme throughout this book. It gives me pain to see it being played out so vividly during our current crisis, but it also galvanises me.

This crisis wasn't there when I started writing this book, and I don't know what state the world will be in while you are reading this book. Whatever the case, I hope the book helps us think more clearly about these contrasting approaches to life, and helps guide us towards a better future.

Part I

GENDERED THINKING

1

Introduction

Being a woman means many things.

Many of those things really have nothing at all to do with being a woman – they are contrived, invented, imposed, conditioned, unnecessary, obstructive, damaging, and the effects are felt by everyone, not just women.

How can mathematical thinking help?

As I am a woman in the male-dominated field of mathematics I am often asked about issues of gender: what it's like being so outnumbered, what I think of supposed gender differences in ability, what I think we should do about gender imbalances, how we can find more role models.

However, for a long time I wasn't interested in these questions. While I was making my way up through the academic hierarchy what interested me was ways of thinking and ways of interacting.

When I finally did start thinking about being a woman, the aspect that struck me was: why had I not felt any need to think about it before? And: how can we get to a place where nobody else needs to think about it either? I dream of a time when we can all think about character instead of gender, have role models based on character instead of gender, and think about the character types in different fields and walks of life instead of the gender balance.

This is rooted in my personal experience as a mathematician but it extends beyond that to all of my experiences, in the workplace beyond mathematics, in general social interactions, and in the world itself, which is still dominated by men, not in sheer number as in the mathematical world, but in concentration of power.

When I was young I didn't have any female mathematician role models. There were hardly any female mathematicians in the first place, but also I felt no particular affinity with those I did encounter, and no special desire to be like them. I did grow up with strong, high-achieving women around me at all levels, though: my mother, my piano teacher, my headmistress, the prime minister, the Queen.

I worked hard to be successful, but that 'success' was one that was defined by society. It was about grades, prestigious universities, tenure. I tried to be successful according to existing structures and a blueprint handed down to me by previous generations of academics.

I was, in a sense, successful: I looked successful. I was, in another sense, not successful: I didn't *feel* successful. I realised that the values marking my apparent 'success' as defined by others were not really my values. So I shifted to finding a way to achieve the things I wanted to achieve according to my values of helping others and contributing to society, rather than according to externally imposed markers of excellence.

In the process I learnt things about being a woman, and things about being a *human* that I had steadfastly ignored before. Things about how we humans are holding ourselves back, individually, interpersonally, structurally, systemically, in the way we think about gender issues.

And the question that always taxes me is: what can I, as a mathematician, contribute? What can I contribute, not just from my experience of life as a mathematician, but from mathematics itself?

What is mathematics?

Most writing about gender is from the point of view of sociology, anthropology, biology, psychology or just out-right feminist theory (or anti-feminism). Statistics are often involved, for better or for worse: statistics of gender ratios in different situations, statistics of supposed gender differences (or a lack thereof) in randomised tests, statistics of different levels of achievement in different cultures.

Where does pure mathematics come into these discussions?

Mathematics is not just about numbers and equations. I have written about this extensively before. Mathematics does *start* with numbers and equations, both historically and in most education systems. But it expands to encompass much more than that, including the study of shapes, patterns, structures, interactions, relationships.

At the heart of all that, pumping the lifeblood of mathematics, is the part of the subject that is a framework for making arguments. This is what holds it all up.

That framework consists of the dual disciplines of abstraction and logic. Abstraction is the process of seeing past surface details in a situation to find its core. Abstraction is a starting point for building logical arguments, as those must work at the level of the core rather than at the level of surface details.

Mathematics uses these dual disciplines to do many things beyond calculating answers and solving problems. It also illuminates deep structures built by ideas and often hidden in their complexity. It is this aspect of mathematics that I believe can make a contribution to addressing the thorny questions around gender, which are really a complex and nebulous set of ideas hiding many things.

How does mathematics do things?

Mathematician Sir Tim Gowers writes of the 'two cultures of mathematics', which he characterises as problem-solving and theory-building.

It seems to me that in the wider world, mathematics is seen as all about problem-solving, and that theory-building is mostly unheard of or otherwise neglected. Of course, the two are not cleanly separate, and in the end I think mathematics is most satisfying when the two come together: where building a theory also solves some problems. But what does it mean to 'build a theory'?

Mathematical theories are descriptive, not prescriptive. They describe something that we see happening at the root of a situation, but they are not only there to predict how the situation will unfold. More broadly, they are there to help us think differently about the situation, to illuminate it, to help us understand how certain aspects of it are functioning. The theory is abstract – we have disregarded some of the surface details, to see what is happening inside.

In mathematics a theory often starts with an idea or a

possibility, and is built up as a collection of conclusions about that idea, or consequences of that idea, or properties of that idea. Sometimes it's really just a reframing of an existing idea, but a slight reframing can lead to a monumentally different theory. If you stand at the top of the building formerly called the Hancock Center in Chicago you can look across the city and see a vast jungle of 'civilisation' sprawling out to the horizon. But if you just shift your angle a tiny bit so that you're looking straight down Michigan Avenue, the jungle becomes a grid system that will suddenly spring up before your eyes in near-perfect alignment. Sometimes in life all it takes is a slight shift in perspective, and this is often how new mathematical theories arise too.

I am going to propose a theory, or possibly just a reframing, to solve a problem.

What is the problem?

The problem I am going to address in this book is the divisiveness of arguments around gender equality. Not all arguments are like this, but too many of them are. Sometimes the arguments push people in opposite directions because of the way we talk about gender, and sometimes arguments are futile because there is little clarity about quite what is being discussed. Take the word 'feminism', for example. One problem right from the start is that there are so many different definitions of feminism – it means completely different things to different people. As a result there is a certain amount of talking at cross purposes. Some people use the narrowest possible definition so that they can obviously justify denigrating it. For example:

Feminism means believing that women are better than men and that men are all bad.

Other people use the broadest possible definition in order to persuade everyone to support it, or even to convince themselves that they are working towards its causes no matter what they're doing. For example:

Feminism means believing that women have as much right as men to choose how to live.

With some people's definition being highly restrictive and some other people's encompassing almost everything, it is a divisive concept before it has even really got past the definitions.

Aside from these arguments about definitions, there are various reasons some people are anxious not to associate with the word 'feminism' at all. Some people stereotype 'feminists' as angry, man-hating, anti-family women, and this is an offputting image for potential feminists of any gender. This highlights another source of the divisiveness of feminism: it inherently seems to separate women from men even as it is trying to overcome that separation. It's hard to argue about men and women without acknowledging that they are different, otherwise we wouldn't refer to them as 'men' and 'women'. And then we get distracted and absorbed by arguments about the ways in which men and women are and aren't different. That argument is a distraction and a detraction.

The problem with this lack of clarity is that it draws us into a meta-argument – an argument about what we should be arguing about. There is an argument that the only experiences of oppression shared by all women are by definition those experienced by the most privileged of women: white women (or, more specifically, rich white straight cisgendered women). There are many problems other than, and possibly worse than, the problems of those women, such as racism, homophobia, wealth inequality. We can thus also be distracted by a competition about which issue is the most important and pressing, instead of addressing them all.

Who benefits from these meta-arguments? I suggest that the answer is: the people who currently hold power and who want to keep it that way. While women fight each other about what feminism should mean, and about which intersectional branch of feminism should be given the loudest voice, and about who is the most oppressed, while men who are oppressed on the grounds of race, sexuality, wealth, status, upbringing, education, gender expression, physical strength or sporting prowess fight to be heard too, while all this disagreement rages between people who feel disadvantaged in any way, the people currently holding power can rub their hands in glee and consolidate that power.

The problem of definitions and meta-arguments is something that mathematics is very good at sorting out.

A mathematical approach

In mathematics we propose a theory by making a basic definition. We illuminate the definition by exploring some key examples. We justify the theory by demonstrating in what way it is ubiquitous and helpful. It might be helpful because it enables us to solve particular problems. It might be helpful because it enables us to think more clearly. I will argue that this one does both.

In maths things need to be ubiquitous *enough* and helpful *enough*. They don't need to be helpful in all possible situations – even being helpful in one situation is enough, although the helpfulness does need to be weighed up against any sense in which it might also be obstructive. This is different from evidence-based arguments. A mathematical theory isn't judged by the statistics of what it achieves in a large randomised sample. When working with evidence we typically test something on a large sample and see if it makes a statistically significant difference. Is it better than doing nothing, on average? In the case of drugs, is it better than placebo, on average? In the case of differences between men and women, are there differences between male behaviour and female behaviour, on average?

Mathematical deductions happen somewhat differently. First of all, they proceed according to logic, not evidence. Secondly, they are more likely to ask: *under what circumstances* does this make a difference? Or even: under what circumstances *might* this make a difference? This is very different from asking whether something makes a difference on average. It is important to remember that averages do not apply to individuals, whether that's the mean, the mode or the median. 'The average person' is not a real person, and saying that 'the average person' does something-or-other certainly doesn't mean that most people exhibit that behaviour.

There are many evidence-based results that might well apply, on average, across a large sample of people, but they do not apply to individuals. Average results based on large samples do not tell us anything concrete about an individual in front of us. This includes the amount of sleep people require (some people really don't require much), the amount of calories needed to maintain a healthy weight (some people need to eat a lot less than the official daily recommended amounts), whether doing exercise makes you feel better (according to research averages it does, but it doesn't work for everyone), whether or not you can win the lottery (it is so unlikely as to be impossible, and yet people do win almost every week).

It is often pointed out that individual experiences do not generalise to large groups, but the reverse is also true: the average experiences of a large group do not apply to individuals. So how are we to think about people?

Mathematical deduction often operates by something more like case study. Something worked in this one situation; what made it work? And can we figure out how to replicate it or get it to work more widely? This is the sort of thought process I will be applying.

Importantly, this is a sense in which talking about personal experience is valid, as long as you're not claiming that your personal experience is necessarily universal, statistically significant, or typical. It's a case study, in which one can ask, as mathematicians do: what made this work, and how can we build on this idea?

I will use many case studies throughout this book, highlighting what I find meaningful about some of the brilliant people who have achieved things in the world. Many of these will be women who achieved things in their own way, not by emulating the heroism, bravery, record-breaking, fearlessness or athleticism of men. Sometimes this means they are undervalued in society and I'll talk about that too. I'll talk about the work of mathematician Emmy Noether, scientists Jocelyn Bell Burnell and Rosalind Franklin, the activist Susan Burton and her advocacy for a more humane rather than punitive justice system, the more famous and less famous achievements of Florence Nightingale, the business principles of Dame Stephanie Shirley and Mary Portas, the career trajectory of Michelle Obama, the experiences of Prof. Deirdre McCloskey in her 'crossing' (as she describes it) from male to female.

As is manifestly evident I am not a biologist, psychologist, philosopher, historian, sociologist, gender theorist, biographer, neuroscientist. I am a mathematician. So I am going to write this as a mathematician, theorise as a mathematician, and explore as a mathematician.

I am going to propose a reframing of the entire discussion around gender. This consists of focusing on relevant character traits instead of gender, and has at its centre some new terminology to help us get started with this new dimension. I will show a vast array of ways in which this can help us move forward, whether we are women or not. Like many mathematical reframings, it is in a way only a very small step further than what many brilliant people have written about and understood before me. But somehow everything seems to stop just short of proposing a new dimension and new vocabulary. Much has been written about the problem without proposing a solution.

My research field, category theory, also began with an apparently small step involving a new dimension and new vocabulary. And yet this had an effect on the entire way of thinking of contemporary mathematics. I think something similar is possible for how we think about gender too. What I am proposing is not a mathematics of gender, but a mathematical approach to gender. I am going to use mathematical thought processes to rethink our whole approach to gender.

The process of maths

Maths is not just about calculating answers. In fact, the theory-building part of maths isn't really about calculating answers at all. Here is a sketch of some of the phases involved when mathematicians build a theory:

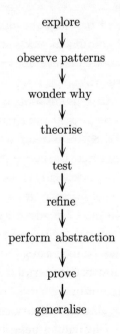

explore
↓
observe patterns
↓
wonder why
↓
theorise
↓
test
↓
refine
↓
perform abstraction
↓
prove
↓
generalise

Often a first step is spotting patterns. At a basic level the patterns might be patterns in numbers, such as the fact that all multiples of 10 end in 0. They might be patterns in shapes, for example in this grid of triangles there are patterns of bigger triangles of various sizes, built up from the small ones:

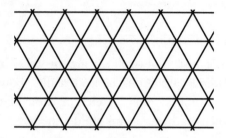

Humans have used patterns throughout the ages and across many (or all) cultures as a way of building up large ideas based on small ones. Nature also uses patterns to build complex structures such as petals on flowers or the spirals of a pineapple. At root, patterns are about connections between different things, or between different parts of the same thing. We can then think of patterns more abstractly to include phenomena that aren't necessarily visual, such as patterns in behaviour, which are essentially similarities between one person's behaviour at different times, or between different people's behaviour, or more broadly between the behaviour of different types of animal, or different communities.

Maths is about making increasingly abstract connections between things. The abstraction consists of forgetting particular details about a situation in order to see some deep similarity. This is how making analogies works, but maths takes it a step further than just declaring that a similarity exists; it moves to a more abstract level by examining what the similarity really consists of, and treating that as a new concept.

For example, the equation

$$1 + 2 = 2 + 1$$

is analogous to the equation

$$2 + 5 = 5 + 2.$$

But in maths we don't just leave it at that – we say

$$a + b = b + a$$

for any numbers a and b. This abstraction makes our point simultaneously less ambiguous and also more open to generalisation, in the sense of broadening to include more examples.

Similarly, if we think about a pattern of women not speaking up in meetings, and female students not asking questions in class, the similarity at an abstract level could be summed up as 'women not speaking up in mixed-gender environments'. At this point we have not measured how prevalent this pattern is, we haven't found what causes it, and we haven't found out how to change it. But in identifying the pattern we have made a start, by shaving off extraneous details and homing in on what is really important. This is a crucial step in building a good theory.

The abstraction then helps as we wonder why a pattern might be arising. Maths is all about asking why, and delving deeper and deeper for more and more fundamental answers. A superficial answer to *why* women are likely to speak up less in mixed groups is 'Because they're women'. We could test that theory, and what we are likely to find is that there are women who actually do speak up, and men who don't. We might find a statistical link between being a woman and not speaking up, but in abstract maths we seek to go a level deeper: *why* does being a woman cause this phenomenon statistically? If it's only statistical then it isn't a fully determined causation, so can we look deeper to find a causation? We might then find that, rather than it being because they are women, it's about relationships between people, and about how different people relate to each other. The idea of studying relationships rather than intrinsic characteristics matches a major advance in modern mathematics: the development of my research field, category theory.

The idea of category theory

Category theory is thought of as a 'foundational' branch of maths because it looks at how maths itself works. Previous to category theory there was set theory, which has a different ideology and thus very different technicalities. Set theory is based on the idea that the fundamental starting point of maths is *membership*, that is, the question of whether or not a given thing is a member of a particular set. This is a bit like parts of society being all about what 'club' you're a member of, whether that is a literal club like the old-fashioned and deliberately exclusive gentlemen's clubs, or more abstract clubs like political 'tribes' of people who believe the same thing, more because of their perceived membership in that tribe than because of anything else.

Category theory takes a different starting point: *relationships*. It is built on the idea that we can understand a lot about something or someone by looking at their relationships with those around them.

Here are some pictures of how those two points of view differ. A set can be depicted as a collection of objects enclosed within a boundary, where inclusion and exclusion are determined by some intrinsic characteristics:

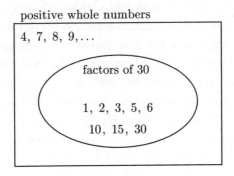

positive whole numbers

4, 7, 8, 9, ...

factors of 30

1, 2, 3, 5, 6
10, 15, 30

By contrast a category can be depicted as a collection of objects with a network of arrows showing how they're related, a bit like in a family tree. In this case, instead of showing parent–child relationships, we can show which numbers are related to each other as factors:

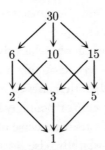

In a way this is only a slight change in perspective, but it took modern mathematics in a whole new direction. For our theory about gender, we can think of the set-theoretic approach as defining the set of women on earth according to some intrinsic characteristics. Sometimes this is cited as being determined by chromosomes, sometimes by reproductive organs, sometimes by hormones. In fact none of these definitions is as clear-cut as some people assume, and furthermore the definitions do not match each other.

I will instead take an approach based in the ideas of category theory. This consists of thinking about how people relate to one another rather than what their biological descriptions are. Across history there have been attempts to define masculinity and femininity in terms of behaviour, and how men and women relate to other people, and there are also definitions of gender as a social construct as opposed to sex as a biological description. But in this book I will instead propose a theory that *only* looks at how people relate to one another, without trying to impose genders on those behaviours. It may or may not be statistically or biologically related to gender, but that would be a different kind of study and is not really the point. Questions of whether certain behaviours are based in biology or society are a distraction; in a way it doesn't matter, it just matters how people actually do relate to one another.

The idea of focusing on relationships *instead* of on intrinsic characteristics is an important part of the idea of thinking 'categorically'. I mean this in the mathematical sense of thinking according to the ideas of category theory, rather than in the non-technical sense of thinking in a fixed, clear-cut, immovable way. It's perhaps an unfortunate reuse of the word because, as I will explain, one of the important aspects of mathematical categorical thinking is flexibility.

The flexibility of category theory comes from thinking about relationships and freeing ourselves from thinking about intrinsic characteristics. This means that we have the potential to apply the ideas much more widely, to situations involving very different types of people, or even to zoom in and out and think about smaller-scale things like games and toys or larger-scale things like schools, companies and society at large. An example I wrote about in my previous book, *The Art of Logic*, involved thinking about power structures giving rise to structural privilege in society such as being rich, or white, or male, or different combinations of them. The relationships then are about the relationship between rich people and non-rich people, or white people and non-white people, or male people and non-male people. All those relationships combine into this diagram showing the relationships between people with different combinations of those particular types of privilege:

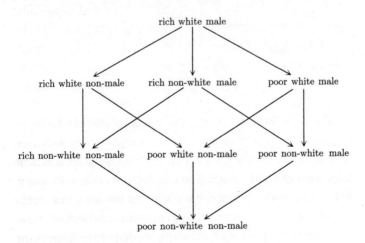

The arrows just depict hypothetical loss of one type of privilege, without making any claims about the causes or consequences.

As we are only thinking about the relationships between people who do and don't have those types of privilege, we can then apply the model to any other type of privilege among any other type of people, such as restricting our attention to women and thinking about the privilege of being rich, white, or cisgendered:

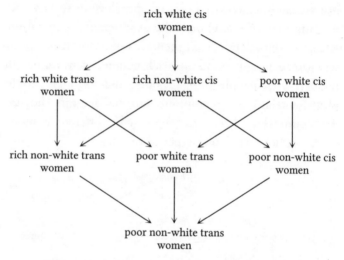

This abstract structure doesn't explain where those relationships come from, or how exactly they manifest themselves in practice, or how to address the situation if, as I do, you believe it needs addressing. But it does enable us to gain some clarity about what the issues are, to focus on the issues that are relevant to a given situation, to package some relationships up into a single unit so that we can hold it in our brain more easily, and to move more smoothly between different situations looking for patterns. It is in this sense that I believe categorical thinking promotes flexibility of thinking, as well as depth of exploration and breadth of application. In fact, this a key point of all abstract mathematics: a flexibility that comes from the act of ignoring some details of a situation, like shedding baggage so that you can travel more smoothly. In maths it enables us to find unlikely similarities between very disparate situations, such as seeing that these diagrams of privilege are the same shape as the diagram of factors of 30 that I drew earlier.

But, as with travelling light, you might worry about what you've left behind. There is a legitimate worry that we will lose something in the process of ignoring some details. Indeed, in arguments in normal life things often do get simplified to the point where important and highly relevant details are lost.

The key in maths is that we choose which details to ignore for now, because of what we have decided to focus on, but that this is a temporary situation. We leave the baggage behind but we don't actually burn it; we recognise that it could be useful for something else. It's a temporary abstraction, in order to explore a particular aspect of a situation, not a be-all-and-end-all abstraction claiming to represent the entire situation.

With regard to gender, this is important because we are going to focus on aspects of the discussion that are only about how people relate to one another, not about intrinsic or biological characteristics; however, we are not claiming that the intrinsic characteristics are never relevant and can be forgotten forever. There are plenty of situations in which gender or sex really do play a role, whether it's because of biological aspects such as reproduction and other medical issues, or because of social aspects including prejudice and harassment, or statistical issues such as average body size and strength.

The idea of a temporary abstraction is to separate issues out and examine them individually, like in a controlled experiment where we try and vary only one thing at a time in order to see what is really causing what. If we conflate gender with character type then we are not doing a controlled experiment, but rather, mixing up issues.

The idea of a temporary abstraction also takes into account the thorny issue of intersectionality, and the need to take other forms of power imbalance into account apart from those governed by gender, for example race, wealth, sexuality, and so on. In abstract maths we don't do experiments but we do study complex structures by trying to examine one aspect of the structure at a time. This is in a way our version of a controlled experiment. The aim isn't to erase the other structures, but rather to see what is really caused by each of those structures.

As a basic example, if we try and study the world of numbers we find that it is a very rich structure in which we can add, subtract, multiply, divide, and more. If we try and study all those things at once straight away, we can become overwhelmed with trying to work out what is really going on. Instead we can study the idea of addition by itself, and see that subtraction is intimately linked with addition so can't really be separated out. Multiplication seems to be linked to addition as it is often thought of as being 'repeated addition' – multiplying 5 by 3 is the same as adding 5 to itself repeatedly: $5 + 5 + 5$. But it can also be taken as a separate operation at a more abstract level, and this gives a wider possibility for application to multiplying things that aren't numbers, such as shapes, sets, or whole structures in category theory.

In this book my approach to gender will be to separate out issues in this way. The first step is to perform a temporary focus on gender issues, even though there are many other pressing issues. The next step is to notice that character traits might be thought of as being linked to gender, but don't need to be, and that we can take them as a separate issue at a more abstract level, giving a wider possibility for application to things other than people, such as structures, systems, communities, arguments. In a sense, this is about introducing a new dimension.

Dimensions

Dimensions, like patterns, can refer to more and less abstract things in maths. At the more concrete level, dimensions are just different independent directions in which we can move in physical space. Our normal world is three-dimensional because there is the north–south direction, the east–west direction, and the up–down direction. (This is mildly complicated by the fact that we live on a sphere, but the principle remains, just with more care needed.) North and south don't count as different dimensions because they're not independent of one another – south is the 'negative' of north, and similarly with east and west. North-east doesn't count as a different dimension because it can be expressed in terms of north and east, so it's not 'independent'.

This idea of an independent new direction gives us a more

abstract concept of direction in maths. The 'direction' can itself be an abstract direction, so that instead of a compass direction for physical motion, it's an ideological direction for thought. If you measure how tall someone is and also what colour their hair is, those things are independent in the sense that you can't state how tall they are in terms of what colour their hair is. Category theory can be thought of as involving one more dimension than set theory because while objects by themselves are considered zero-dimensionsal, relationships between them make connections, like paths, which are one-dimensional. One example of this is what happens when we put numbers on a number line: they go from a collection of random zero-dimensional points into a one-dimensional line built from the relationships between the numbers:

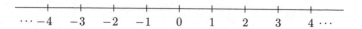

We are going to introduce a new dimension to the discussions about gender, and that dimension is to do with the ways people relate to one another. I am going to think about ways of relating that I believe are particularly relevant to the inclusion of women in society, but we will approach it in a way that doesn't need to be directly tied to gender. Not only is it a different dimension from gender, I believe it is a different dimension from the existing ways in which we think and talk about character traits.

It is hard to define a new dimension because you can't express it in terms of the existing ones. However, this form of abstract invention is prevalent in maths. It might seem a bit like making things up, but that doesn't have to be a bad thing. One example illustrating this is when mathematicians invented a new type of number called *imaginary*, by taking square roots of negative numbers. We sometimes say you can't take the square root of a negative number, but really we just mean that no ordinary number can be the square root of a negative number. So what about some other kind of non-ordinary number?

This is how mathematicians came up with the vividly named 'imaginary numbers'. We basically just invent a number i that is going to be the square root of -1, so $i^2 = -1$. We don't know what i 'is', we just know its relationship with -1. We can also conclude that if the ordinary numbers go on a line, this funny new number i can't be anywhere on this line. So we might as well depict it in some new, independent direction, like this:

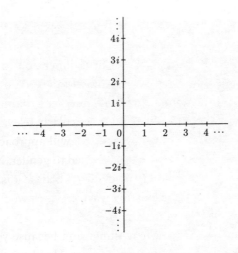

This is a sense in which we have invented a new dimension. You might think we just made it up, and moreover, we even acknowledge this by calling it 'imaginary'. But we've made up everything in abstract maths really. We've made up the ordinary numbers 1, 2, 3, and so on – what in fact are those? They are just ideas. So we can make up the idea *i* too. We can make up any idea as long as it doesn't cause a contradiction.

The extraordinary thing is that this apparently fictional world turns out to shed a lot of light on our ordinary, familiar world. Sometimes it turns out that we have been looking at the world in too low a dimension to understand it, like if you look at shadows on a wall. Those shadows are a two-dimensional projection of a three-dimensional scenario, and it can sometimes be very hard to work out what the real three-dimensional situation is. In fact, sometimes the shadows can mislead us into thinking something else is going on entirely, such as in this picture:

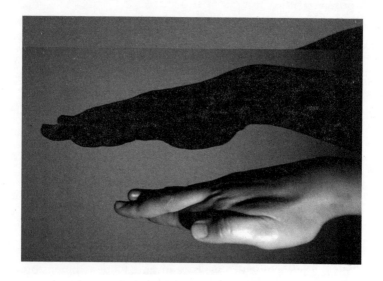

Misleading 'shadows' also occur with abstract dimensions, including the one I will introduce to rethink gender. We have been examining questions of gender inequality without this dimension, and I believe that this has been leading us astray; I believe that this new dimension, even though I sort of just made it up, sheds light on familiar situations. In fact, like category theory after set theory, this is in some ways only a slight change in perspective, but my belief is that it's a slight shift that can radically change how we see things, adding radical clarity and simplification of the best kind.

New dimensions are hard to work with because they're new, but it can be remarkably helpful in maths to name them. The act of naming is another small step, but can quickly help us organise our thoughts and then go further with them. New terminology is an idea that has been proposed by other authors on this subject, but I am going to go a step further and actually propose some words. You can have the thoughts without the words; you can also have the thoughts with different names. But it would have been hard to develop category theory without the word 'category' for the new type of structure we're thinking about; it would be hard to study the made-up square root of -1 if we had to keep referring to it in that long-winded way instead of by its new name i, and I think it is hard to proceed with our arguments on gender while we're stuck with our old and often gendered terminology. Naming a new theory is a starting point for making progress with it.

How we make progress

The type of progress we can make with this new theory is

broadly twofold, and can be evaluated in these two ways. When we evaluate a theory in maths there is a practical side, that is, what the theory enables us to do, and a theoretical side, that is, the generality and flexibility of the theory, and its ability to balance finding a broad range of examples while also still separating out pertinent issues. The practical and theoretical are linked, of course, and the theory behind the theoretical evaluation (if that's not too convoluted) is probably that these are signs of a theory that will eventually enable us to do many things, even things that we don't know about yet.

One of the important things category theory enables us to do is develop more nuanced ways to think about sameness. If we think about intrinsic characteristics, then we are often stuck saying things are the same if they have the same intrinsic characteristics. But if we move to talking about relationships, we can allow for things and people who relate to others in the same ways even if they have different intrinsic characteristics. In some cases intrinsic characteristics might be important, such as if someone is a body double for someone in a film – but even then they only need to be similar in certain terms: for example, the male stuntman Gary Connery doubled for the Queen in the opening ceremony of the 2012 Olympics when the Queen was shown apparently jumping out of a helicopter and arriving by parachute.

When discussing gender issues, we can get stuck in situations where we attempt to say that men and women are 'the same' while also trying to address the fact that there are far more men than women in many walks of life such as maths, science, business, politics, wealth. Some people simply shrug and insist that men and women are different, so this imbalance is not unfair. Others agree that this situation is bad (because men and women are 'the same') but still think we can't really do anything about it, because we must treat men and women 'the same'.

The categorically inspired approach will show us ways in which we can treat men and women the same if they relate to others in the same way. This doesn't mean that all men and women are the same, which is absurdly oversimplistic. It means that we can find the types of behaviour that are important or beneficial, find people who exhibit that behaviour, and treat them as 'the same'. For example, instead of saying something like 'men ask for pay rises more than women and that's why they're paid more', we could say 'people who ask for pay rises are paid more' (regardless of whether they're men or women).[1] We can go one step further and then ask whether this is actually how we want the system to work: should people be paid more because they ask for more? I believe they should be paid more according to evaluation of their actual work, not just their ability to ask for a higher salary. This provides a more nuanced solution although it is also more difficult than simply paying women the same because 'men and women are the same'. But it is crucially less divisive.

The idea of sameness can be applied to individuals inside structures and also to structures themselves. This is a form of zooming in and out that is very typical of the sort of flexibility that category theory – and abstraction in general – gives us. Concrete things are too fixed to be used in this

---

1 It is worth noting that there is evidence that women aren't paid as much even when they do ask for pay rises.

way. A book is a book, but if we go more abstract and think of a book as a collection of ideas communicated with some unifying coherence, then we can zoom out and apply it to libraries, and we can also zoom in and apply it to individual chapters or even individual paragraphs.

The new dimension that we use to help us break out of the confines of gendered thinking will have to be abstract enough to be applied to people, but also applicable to collections of people, or processes. We can apply it to activities, relationships, teaching styles, even to different types of maths. This provides a benefit of this level of abstraction that I don't believe we can get from any of the existing words describing character traits that might somehow be related to our new dimension.

This is one of the theoretical ways in which a theory is deemed good in maths, and is one of the ways in which a theory has the potential to grow, as it can be applied so much more widely than originally conceived. Take the idea of symmetry, for example: it starts as something that we can see in objects that can be folded in half with both sides then lining up:

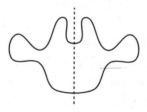

But if we think of symmetry more generally as a relationship between an object and itself in which we can somehow move the object and it still looks the same, then we include the idea of rotational symmetry like windmills, or this shape which is made of the same halves as the one above, but rotated rather than reflected:

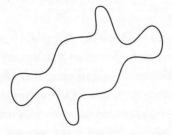

If we expand this to other types of transformation on objects other than just moving them, we get the idea of something that looks the same as itself if we zoom in or out, which gives us self-symmetry such as in this fractal tree:

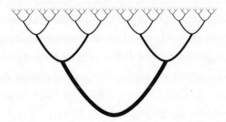

If we expand this further to include abstract transformations (as opposed to just physical ones) we get symmetry such as in this formula:

$$a^2 + ab + b^2.$$

This has abstract symmetry because if we transform it by swapping a and b it becomes

$$b^2 + ba + a^2,$$

which is in fact the same as the original.

This is why a new theory in maths can never be comprehensive at the start – or rather, if it is, then it must be a rather limited theory. A good theory has unlimited scope for expansion, just as category theory was originally applied to a very specific part of abstract algebra and then grew to include the rest of pure maths, then theoretical physics and theoretical computer science, and more recently even biology, chemistry, engineering and business.

Category theory has grown in this way by abstraction, flexibility of thinking, and also by looking at things from different directions: from the outside in, and from the inside out. Sometimes we pick which relationships we want to study and see what world this makes, but sometimes we think about the kind of world we want to study and see what relationships will create this world for us. It's a bit like the internal and external approaches to maths and indeed life in general, where you can pick a path and see where it takes you, or you can pick a destination and then examine which paths will take you there.

If the two agree it's usually a sign that you're onto something good, and it can also be enormously satisfying. If the two disagree then you need to decide which one you care about more. If you're trying to visit the Eiffel Tower but you see a very beautiful street that looks enticing, do you turn down it and risk missing the Eiffel Tower or do you forge ahead and forgo the beautiful side street?

Abstract maths is different from the concrete world because the things we're looking for can be dreams or dream worlds.

Dream worlds

A fantasy world isn't exactly 'real' (whatever that means) but can still help us shed light on the real world around us, just as fiction can (which is why it's so important to take interpretative art seriously). Dreaming about the sort of world we want to see can help us sort out what our fundamental beliefs are, as can zooming in and dreaming about the sort of life we want to live. For some people this means unlimited riches in order to own designer clothes, a private jet and a yacht, but for others, including me, it means helping others understand things with the general aim of lifting up the disadvantaged. Those aims are rather different, and entail rather different activities to achieve them. However, dreaming about what I would do if I did have unlimited riches helped me understand my fundamental beliefs and thus make changes in my life that I believe are beneficial both to me and to the world, even though I don't in fact have unlimited riches.

Maths isn't just about getting the right answers but is about dreaming up different worlds in which different things can be true. For example, children (and adults) often ask if infinity is a number or not. The answer is that it depends on what world of numbers you dream up. Infinity is not in our ordinary world of everyday numbers, and if we just try and throw it in by 'inventing' it naively, we end up with all sorts of contradictions and paradoxes. But there are many ways to dream up different worlds of numbers that can include infinity without causing those problems.

In category theory we can dream up worlds in different ways. We can decide what type of relationship we want to focus on, and see what world this produces. But we can also decide what kind of world we want to see, and then look for the kinds of relationships that will make that happen. Seeking social justice, in whatever form we believe in, can also be helped by both of those approaches, and in this book I'll do both of those things. We can benefit from dreaming up our imagined utopia in order to help us understand the overall features we really care about, and then we can try and work out what small steps will take us there from our current situation. But we can also zoom in and think about what sort of interactions we care about between people, and then patch those localised features together to see what global structure will result. Sometimes problems arise because people hold fast to the local features and disregard the resulting global ones, such as insisting on only evaluating people by their achievements (not taking any life disadvantages into account such as gender, race or socio-economic background) and disregarding the fact that this perpetuates the under-participation of those disadvantaged groups in society. If local behaviour that seems fair results in global outcomes that are not fair, I believe that rather than just disregarding the global consequences, we need to make a conscious choice about how much that matters to us, like with the Eiffel Tower and the enticing back street. If dreaming about utopia and thinking about local issues take us to the same global outcome, then so much the better. I believe this to be true of the theory I am proposing.

I will end the book with my dreams for a world enlightened by these new ways of thinking about gender. I must acknowledge that for many people gender bias is not the most life-threatening problem they face. But I believe it is still a serious problem in the world, despite all the wonderful work of people who have gone before me. And I believe in trying to help the world in ways that we can, while recognising our limits and always trying to do better at working towards the world we want to see.

I want to see a fairer world, in the sense of people not being held back by features about themselves that should not be relevant. I think that gender is often relevant in practice when it shouldn't be, and I dream of a utopia in which we focus on relevant character traits instead of gender, and in which we do not associate those traits with genders as we have so often done historically – for example with traits like strength, ambition, confidence, empathy, kindness, communication skills. I don't want to see men and women pitted against each other in a battle for dominance. I don't want to see 'masculine' and 'feminine' character traits evaluated against each other in a zero-sum game in which only one side can win, and in order to do so the other side has to lose. I don't want to see women urged to be more like traditional men in order to succeed, when there are other ways to achieve the same success, and indeed other ways to make worthwhile contributions to society instead of those traditional forms of success.

I want to see more unified discussions and unifying solutions to these issues, where we avoid separating men from women and erasing non-binary people in the process. The world isn't a dichotomy between genders. The world consists of all of us together. This book is my dream of how we can use mathematical thinking to build a better, fairer world, and the beginnings of a plan for how we might be able to get there, starting small and building up gradually to encompass and reform more and more of our flawed social structures.

In two-dimensional maths we draw two-dimensional graphs of *y* against *x*. But *y* doesn't have to be pitted against *x*: *x* and *y* can be added together to make a new dimension and a new way of thinking.

2

The difficulties of difference

Are men and women innately different in some way? And if so, is it justifiable to treat them differently?

We are going to begin by examining these related questions and, more importantly, the sorts of arguments that are used to justify answering yes. We are going to examine and refute these arguments mathematically. We're going to show the weakness of arguments that suggest gender imbalance is 'just the way of things'. But in the end, rather than just refuting these arguments, we need to reframe the entire debate so that we can stop thinking about gender differences where they aren't relevant, and stop getting involved in arguments that mainly serve the people who currently hold power in society.

Why do we persist in thinking in terms of gender differences? It's telling to consider who benefits, and question why this research is even being done. Why is anyone trying to prove that there are innate differences between men and women in intelligence, scientific ability, competitiveness, or any other traits that seem to confer high status in society? One general reason to cling to the idea of innate ability is to give ourselves an excuse for not being good at something. If I claim that I just have no natural aptitude for sport, that gives me an excuse for being really very very bad at sport. Conversely, when people declare that I am very 'talented' at the piano, that negates the thousands of hours of practice I have put in. People can declare themselves to be a right-brained 'creative' person, and use it as an excuse for being disorganised. They can boast of being a left-brained 'logical' person as an excuse for being insensitive. (This is in spite of the fact that the left/right brain theory has been largely debunked.)

The more invidious reason to claim that people are born with certain traits is to avoid having to help them do any better. This is a way of not having to address our prejudices. If we can somehow argue that women are innately less intelligent than men, then we won't have to address issues of inequality in education, science, business, politics and every echelon of power. If 'innate' biological differences are found, they are like cannon fodder for people who seek a pseudo-rational basis to maintain structures that discriminate against women.

If the arguments are about biology, what can maths do for us here? Maths provides us with a framework for making justifications and also for evaluating them, so it gives us a way of assessing the value of any particular opinion. This is why maths can be relevant to all sorts of things that don't appear to be obviously 'mathematical'. Mathematics is too often seen as being all about numbers and equations, in which case anything that does not involve numbers or equations appears to be not 'mathematical'. But I think that anything that involves some sort of justification can be examined mathematically.

A mathematical justification is called a proof. It is like a kind of journey. It has a starting point, a destination, and a way of getting from the starting point to the destination using logical deductions. And so we evaluate it by thinking about the starting point, and thinking about the logical deductions.

In addition, a journey is often not just about getting there, but about what you can do when you get there. Sometimes you can't do much except admire the view and come back, say if you climbed a mountain. But at other times you arrive in a city rich in culture where you can explore, learn, and challenge your world view.

We are going to use this approach to evaluate some existing arguments about gender differences, and then make a theory of how these arguments are flawed. However, these existing arguments are not stated quite like mathematical proofs, and so the first thing to do is to find the (attempted) logical structure of the argument and express it a bit more like a mathematical proof by stripping it to its bare bones. This process of peeling away outer layers is an important step in the mathematical process. The outer layers often obscure what the real structure of the argument is, a bit like sleight of hand, and so removing those layers often exposes the flaws in the argument. This is one of the reasons why maths uses very precise language and abstractions, to leave less possibility for that sort of misdirection. It's a bit like the fact that it would be hard to carry a concealed weapon on a nude beach.

Case study

Here is one much-discussed argument about the gender

imbalance in science and maths, involving the idea of assessing people according to 'systemising' and 'empathising':[1]

> Men's brains tend to be stronger in systemising than empathising, and systemising is important in mathematics, so it is to be expected that there are more men than women mathematicians.

This looks a bit like a simple string of implications:

1. Being a man implies being better at systemising.
2. Being better at systemising implies being better at maths.
3. Therefore being a man implies being better at maths.

Now, if these were valid logical implications of the sort used in mathematical proofs, then the conclusion would be correct. This is because in pure logic if we know 'X implies Y' and also 'Y implies Z' then it is logically valid to conclude 'X implies Z'. This sort of inference is exactly how complex logical proofs are built up from small steps, where the steps all neatly fit together to form a continuous staircase from the starting point to the conclusion.

However, in the situation above, they're not really *logical* implications. They are something more complex and difficult. The first step is a statistical observation, not a logical

---

1 J. Billington et al., 'Cognitive style predicts entry into physical sciences and humanities: Questionnaire and performance tests of empathy and systemizing', *Learning and Individual Differences*, vol. 17 (2007), pp. 260–68.

implication. It has been observed that men, on average, tend to be better at systemising than empathising, according to some proposed definition of these things. The next step, the idea that systemising is important in mathematics, has a status that is somewhere between an assumption and an observation. The idea that it is important in mathematics sounds logical, but that makes some assumptions about what 'systemising' really means, and what skills are really important for research mathematicians (as opposed to people who are very good at mental arithmetic or maths exams). There are some observational studies backing up this idea, but in that case the result goes back to being an observed statistical correlation.

The fact that these are statistical observations then raises the question of whether the effect is something innate about men, or something cultural. A more honest chain of argument would go like this:

1. Men have been observed to be statistically more likely to be stronger at systemising than empathising, in some very specific definitions of these words.
2. A correlation has been found between this notion of systemising and becoming a mathematician.
3. Therefore we might expect more men than women to become mathematicians.

This is a rather weaker conclusion, reflecting how weak the steps in the argument actually are. It tells us nothing about whether or not it is fair or biologically inevitable that the gender imbalance persists.

I am now going to propose a general theory of these weak arguments, based on the above blueprint.

A theory of weak arguments

One important step in the mathematical process is to make a general theory that can then shed light on more than just one situation. We often do this with the help of abstraction, stripping away some external details to show the bare bones of a situation, which can then be seen as the bare-bones structure of other situations. This was the point of my introducing the letters X, Y and Z in place of some parts of the argument above – to focus on the logical structure of the argument that did not really depend on the details of what X, Y and Z actually represented in this particular case.

Having done that to show what a sound logical argument would look like, we can constrast it with what the weak unsound argument looks like, which is something like this:

1. Men are observed to have quality Y on average, under some select circumstances.
2. Quality Y is believed to be good for activity Z without any very strong basis.
3. 'Therefore' men are naturally better (or worse) at Z.
4. 'Therefore' we don't need to do anything about the imbalances in favour of men in activity Z.

We will be focusing on issues of gender in this book, but it's worth noting that this general argument form is applicable very widely to many situations other than gender where arguments about imbalance are raging, including disagreements about race, wealth, educational background, sexuality, and so on. One advantage of abstraction is that it helps us to see connections between a broad range of situations beyond the matter directly under consideration.

Anyway, the weak argument gets subtly but invalidly morphed into one that seems much stronger via a series of sneaky slides as in the above example. 'Men are statistically more likely to be better at sytematising than empathising' turned into 'Being a man implies being better at systemising', involving some unsound deductions about statistics. The abstract version of this slide is something like this:

men have quality Y on average
↓
men have quality Y

There's another slide that turns 'Men are observed to be better at systemising' into 'Men are by nature better at systemising', assuming the effect to be by nature not nurture. This is the sort of deceptive argument that enables some people to assert that gender differences are biological and therefore not the fault of discrimination. The abstract version is like this:

men are observed to have quality Y
↓
men naturally have quality Y

And then there's the slide that turns 'Men are better at systemising' into 'Men are better at maths', where the thing that has (supposedly) been measured is taken as a proxy for something much harder to measure. The abstract version is like this:

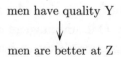

men have quality Y
↓
men are better at Z

where Y has been casually swapped for Z without much justification or fanfare. These three surreptitious 'slides' can be combined to make arguments dramatically weaker by means of such less noticeable increments. This means that where we start at the top of the following diagram, we can sneakily claim that we are anywhere further down it by sliding down the arrows, but each time we move along an arrow the argument becomes more flawed:

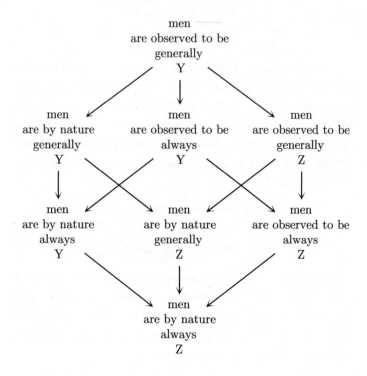

The generality of this theory means that it can be applied to a wide range of examples where gender imbalance is found. In mathematics a theory is judged by the breadth of examples it unifies and the amount of light it sheds on those examples, so after making a mathematical theory we typically test it by trying it out on some more examples. We could try applying it to another type of argument that has been used to justify gender imbalances in academia, this time in physics:[2]

1. Men have more academic citations than women in physics.
2. Citations are a measure of how good you are at physics.
3. Therefore men are better than women at physics.
4. Therefore it is fair that there are more men than women in physics.

The first point is fairly well documented, but the second assertion involves less of a slide and more of an enormous leap of faith. The conclusion that 'men are better than women at physics' may well be true, statistically, if we take a snapshot in time right now and take 'better at physics' to mean more successful at making progress in advancing theories, but concluding that this is a fair situation is another giant unjustified leap: they might be more successful because the world favours them unfairly. In this case the issues were less about the observations being taken and more about the conclusions. We will

---

2 See, for example: A. Strumia, 'Gender issues in fundamental physics: A bibliometric analysis', preprint 2019.

discuss those issues, relating to the bottom-right part of the diagram, in the next chapter, and focus on the top-left part of the diagram for the rest of this chapter: the part of the argument that is supposed to show us that there are some innate differences between men and women.

There is a strong perception of differences between men and women, and understandably so – there are some fairly obvious general differences between men and women physically. But we are going to discuss the flaws of taking those differences too seriously, or concluding too much about those differences. Instead of asking whether or not gender differences are innate, it is more productive to ask in what sense they are innate, to what extent they are innate, and what the point is of basing our world on those differences.

When it comes to making arguments about gender differences, we have seen some flaws in the logic used above. There are also flaws of method, for example, in how we measure our observations in the first place, and flaws in how we understand the statistics even if some robust observations have been made. We will start at the beginning with the observations.

Problems with our observations

Sometimes our conclusions based on observations go wrong because we just base them on our own experience. This could be unsound because we have a small sample which is skewed by our own situation. For example, I happen to know an unusually large number of women mathematicians.

It is even more unsound if we have succumbed to confirmation bias, where we tend only to notice evidence that supports our theory and not evidence that refutes it. This common type of bias is well known and well documented, but many people who seem to know about it intellectually still fall prey to it in practice. A male maths professor saw me checking my hood in a mirror when we were all lining up for graduation and said, 'See, women are just vain.'

Where personal observations are prone to these sorts of errors, science is supposed to be based on a much more rigorous and unbiased process of observation. Experiments are controlled, sample sizes are large, and the replicability of scientific experiments together with peer review are supposed to ensure that the observations are impartial and reliable. Unfortunately this has not been the case when it comes to the search for differences between men and women. The prejudice-based flaws have been analysed very comprehensively in books such as *Inferior* by Angela Saini, and *Testosterone Rex* by Cordelia Fine. Note that this does not mean that science itself is flawed – sometimes people see any flaw in a scientific result as evidence that science is not 'trustworthy' and we might as well go back to relying on personal opinion. The scientific process is a process, and part of that process is a process for finding flaws, and the fact that scientists are able to find flaws in scientific work is a sign that the process is working.

I will focus on the flaws that I can illuminate by means of logic and abstraction, rather than on the flaws in the experimental methodology. I will begin with the issue of trying to take observations to show 'innate' properties of people in the first place. One rather serious problem is how to ensure that any differences you're observing really are innate and not learnt.

Imagine you are trying to design an experiment to

investigate the innate differences between men's and women's brains. How are you going to separate out nature from nurture? One way to do this is by studying identical twins who were raised in completely different environments. However, if we're trying to investigate, say, mathematical ability, then we would need to find a pair of identical twins raised in different environments where one became very good at maths and the other didn't. Or where they both did, or both didn't, despite the differences in upbringing. This is a little far-fetched, even aside from the slight problem that identical twins will normally be the same gender.

The other popular approach is to try and do the research on children when they're young, so young that socialisation and environment have not yet had an effect. The trouble is that those things have an effect almost immediately, so the studies have to be done on babies who are just days old. (People often interact with babies differently if they're told they're a boy or a girl, right from the start.)

The next problem is trying to make a controlled environment for a baby. You'd probably have to get a robot to interact with the baby rather than a human. Plus, how is it going to be double-blind with respect to the gender of the baby? This means that the people doing the tests and recording the data must not know the genders of the babies. You would have to take the newborn babies from their parents, leave them all without gender identifiers in a separate room, and have the researcher come and collect the babies without interacting with the people who had taken them from their parents. They would then take them in turn to a room where they would be tested by a robot. Now try persuading any parent to let their newborn baby take part in this study.

The next problem, if you've managed to prise some newborn babies from their parents, is: what tests can you do on newborn babies? You can't very well ask them to solve maths problems. Thus the experiment has to test something extremely basic, usually just timing how long different babies look at different pictures.

If any differences are measured, then the experiment will tell us exactly this: that some babies looked at some pictures for longer than other pictures. In one much-cited experiment[3] the babies were looking at a human face or a mobile. In another[4] it was pictures of a car as opposed to pictures of a doll.

What are we then to infer about men's and women's brains? There is a rather large leap from babies looking at pictures for a few seconds longer, to the gender imbalance in maths professors at research universities. And yet this is the sort of inference that happens:

1. Girl babies looked at the face for slightly longer than the boy babies on average.
2. Therefore girls' brains are hardwired for empathy whereas boys' brains are hardwired for systemising.
3. Maths is about systemising: therefore this shows that men are hardwired to be better at mathematics.

---

3 J. Connellan et al., 'Sex differences in human neonatal social perception', *Infant Behavior and Development*, vol. 23, no. 1 (2000), pp. 113–18.
4 V. Jadva, M. Hines, and S. Golombok, 'Infants' preferences for toys, colors, and shapes: Sex differences and similarities', *Archives of Sexual Behavior*, vol. 39, no. 6 (2010), pp. 1261–73.

This involves sliding through all the sneaky ways to subtly change an argument, with one part going through an extra step:

boy babies looked at the mobile for slightly longer

↓

male brains are hardwired for systematising

↓

men are hardwired to be better at mathematics

The above problems are about what we infer from the behaviour of babies. There is essentially a trade-off between the certainty that what we are studying really is innate and not the result of 'nurture', and the certainty that what we're studying is actually linked with some meaningful aspect of adult behaviour. This trade-off is summed up in this diagram:

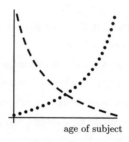

age of subject

certainty that what we are measuring is:
- − − nature, not nurture
- • • • indicative of adult behaviour

Aside from this tricky question of separating nature from nurture, there is also the difficulty of actually finding something quantifiable to measure. If we are trying to study the intelligence of men and women, how are we going to measure intelligence? Tests that produce clear answers are necessarily very restrictive and thus only test something extremely restricted, and therefore the question of how this relates to whatever we actually mean by 'intelligence' remains a big one. IQ tests only test your ability to take an IQ test (on that particular day). Historically brains were weighed as a supposed way of studying the relative intelligence of men and women. This measure is certainly unambiguous and objective, but using it to indicate intelligence presupposes a link between brain weight and intelligence. In fact, historically, scientists assumed that men were more intelligent than women, found that men's brains weighed more on average, concluded that brain weight must determine intelligence, and then used the fact that men's brains weigh more to conclude that men are more intelligent than women. It's a breathtakingly circular argument.

More specifically, we might try to study differences in mathematical ability in order to decide whether or not it is 'fair' that there are so many more male maths professors than female ones. But how do we measure maths ability? If we use tests of mental arithmetic then all we're testing is the ability to do mental arithmetic, and the question remains of how, if at all, that is linked to one's ability to do research maths. (Plenty of non-mathematicians are better at mental arithmetic than I am, especially those who use it every day, which I do not.) Attempts to link mathematical ability to the 'systemising brain' use a questionnaire with questions such as 'If you see a mountain do you think about the geological processes that formed it?' and 'If you have an electrical problem at home can you fix it yourself?' The answer for me is no to both, but all that says about me is how I feel about mountains and electricity, not how good I am at maths. Even more tenuously, the questionnaire asks about my understanding of people in social situations, and then seems to conclude that if I am able to understand people then this is an indication that I am bad at maths, which is an egregious myth to be perpetuating.

This is one of the flaws in the study that attempted to assess physics ability in men and women by measuring the number of citations of their papers. Number of citations is definitely measurable, but the link between that and how good a physicist is is tenuous, and is affected by the implicit bias that may cause people to cite male authors more than they cite female ones.

In short, there is a trade-off between how measurable something is, and how meaningful it is likely to be in terms of actual behaviour. This is summed up in this diagram:

Aside from these essential logical problems with trying to study gender differences by observation, there are problems even if the observations we made were sound. Inferring things from observed behaviour differences is fraught with problems, many of which are to do with misunderstandings of how statistics work. In the rest of this chapter we'll explore the ways in which statistical results are simplified in order to summarise them briefly, losing crucial nuance in the process. Unfortunately the modern world is more about click-baitery than nuance, and so a headline stating that 'men are better than women' at something is going to be favoured over one stating that the distribution for men is a little further up than the distribution for women but that there is a large overlap. One thing that we readers can do is educate ourselves about this and other nuances so that we don't fall for these black-and-white melodramatic headlines any more.

These issues will be familiar to readers fluent in different types of average, standard deviation and shapes of distribution, but as these are a widely used method of misinformation I have included some demystification here.

Averages

If we're trying to make a sweeping statement about men being different from women, it might seem that we can put ourselves on a more secure footing by saying 'on average'. But declaring things are true 'on average' can still leave many possibilities for individual cases. The relationship between the 'average' features of men and women and the behaviours of individual men and women is complex, and knowledge about individuals can't be reconstructed from just knowing the average. Men may well be faster runners than women on average, but that doesn't mean that every man is faster than every woman.

Averages and percentages conjure up traumatic memories of school maths classes in many people, and unfortunately this makes them an easy tool for misinformation and misrep-resentation. Some unscrupulous people do that deliberately to manipulate public opinion, while others might be well-meaning but misguided. The result is that statistics can be used to make people seem more different than they are, but greater mathematical rigour can help us see through that.

When we're thinking about a large collection of data we usually need some way to sum it up more succinctly so that it's not just a huge list of numbers. We want to be able to understand its general shape and compare it with other sets of data. There is always a trade-off between succinctness and loss of information. The more succinct your summary is, the easier it is to present and grasp, but also the more information you will have lost in the process of making the summary.

Averages are a way of summarising a collection of numbers to just one number: we are doomed to lose huge amounts of information and nuance in the process. Sometimes that information is critically important, and hiding it is a way to manipulate information in favour of a particular point of view. Different types of average focus on different aspects of the information, and hide the rest.

Here is an example, admittedly an extreme one, to illustrate the point. The figures listed are some hypothetical salaries of this hypothetical company employing five men and five women. The type of average called the mean is found by adding all the salaries together and dividing by the number of people; in this case I have calculated the mean for men and the mean for women separately:

	men	women
	£200,000	£10,000
	£200,000	£10,000
	£200,000	£10,000
	£200,000	£10,000
	£200,000	£1 million
mean	£200,000	£208,000

This company has one very highly paid woman, which skews the mean salary for women so that it is higher than the mean salary for men. The company might well boast that 'it pays women more than men on average', or that 'the average woman at the company is paid more than the average man'. They might further shorten this to the more eye-catching 'women are paid more than men' in a linguistic sleight of hand.

I hope that hearing these descriptions and looking at some figures gives you at least a feeling of queasiness that something here is not right. I have seen cases like this where the data is produced, the statistics are analysed, the description is made, and the case is declared closed: gender pay equity has been reached, and in fact it's now unfair because women are paid more. The women feel hard done by, as they largely know that they are underpaid, but they also feel helpless because the 'hard data' does not seem to back them up.

When I feel this sort of queasiness, it is a signal to me that something is amiss in the logic somewhere. It is then the job of my brain to work out exactly, *precisely* what is causing that unease, almost like a doctor using symptoms to pin down a diagnosis. If I find the root cause precisely enough, then I can fix it, and the queasy feeling goes away.

In our hypothetical gender pay example, escalation of phrasing went like this, in small steps:

the mean salary for women is higher than the mean salary for men

↓

the average salary for women is higher than the average salary for men

↓

women are paid more than men on average

↓

women are paid more than men

It is popularly quipped that there are 'lies, damned lies, and statistics', but it really isn't the statistics that are lying here: it is the interpretation or presentation of them. Once we understand how flawed the mean is as a measure, then we need not be fooled by the rhetoric accompanying these statistics any more. In fact, it's not a bad idea to be highly suspicious any time anyone mentions an 'average' in any context. There: the reaction of horror felt by those traumatised by averages at school is in this sense quite sensible.

In the above example the problem arose from the distribution of salaries for women being highly asymmetrical, and the discrepancy can be revealed by considering medians instead of means. The median is found by listing the numbers in order and finding the middle entry. The median salary for men is £200,000 whereas the median for women is just £10,000 – so much for equity.

While the supposed 'equity' in this case has easily been debunked, it is not always so easy. Here is a more nuanced example in which you might feel in your gut that women are worse off, yet the mean and the median are both higher for women than for men.

	men	women
	£100,000	£10,000
	£100,000	£10,000
	£100,000	£105,000
	£450,000	£105,000
	£450,000	£1 million
mean	£240,000	£246,000
median	£100,000	£105,000

The problem here can be detected by also looking at the 75th percentile and the 25th percentile, but I hope you can see (or guess) by now that I could construct a data set in which women were paid more according to all these measures but they're still in some sense worse off. The more percentile points you consider, the harder it is to hide your inequalities, but after a certain point you really might as well look at the shape of the whole distribution.

Shapes of distributions

The picture of a whole distribution can give us a much better overview of a situation. Here are some examples where in theory the mean could be the same in every case, but the way the data clusters or spreads is very different.

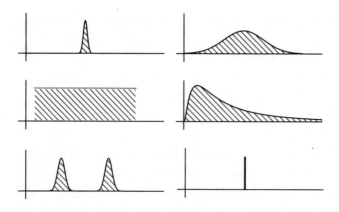

There is no type of average that tells us everyone behaves according to the average. That could only happen in extreme situations such as in the picture that just looks like a spike. There isn't even a type of average that means *most* people behave in that way. We have looked at the idea of mean and median, and it's important to remember that the mean doesn't represent anything very realistic in real terms except in select circumstances. If we're talking about the amounts that people are paid, as we did above, this would tell us how much everyone would be paid if we took the pot of money and divided it between everyone equally. But with something like height or 'length of time spent staring at a picture' that's not very meaningful. However, the median does have a real meaning: that exactly half the results were bigger than this and half were smaller.

The mean will be in the middle if the distribution of data is symmetrical, like a normal distribution shown below on the left. But on the right is an example where the mean is very far from the midpoint, which can happen if a few very extreme figures skew the results in one direction – as in our earlier example of the company with one very highly paid woman.

This last example is called the 'log-normal' distribution (the log of the results has a normal distribution). This could be a graph of wealth distribution, for example. The super-rich people skew the mean towards the right, but far more people have less wealth, which is why the peak is towards the left.

Many distributions of data are symmetrical, and in fact many have the sensible 'bell curve' – the normal distribution, as depicted above, that peaks in the middle and tails off evenly to both sides. In that case not only is the mean the midpoint of the distribution, it also indicates a clustering so that most results really are clustered around that mean; but with different shapes of distribution the most common result could happen somewhere else entirely. The result that appears most often is called the mode. For the log-normal distribution, we can see how the skewing results in the mean, median and mode being quite different:

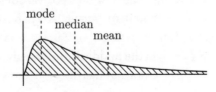

The mode is not very helpful where there is a very even distribution with roughly the same number of people achieving all results; in a situation with two peaks none of those averages is at all useful:

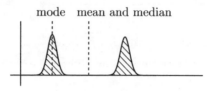

This is called a bimodal distribution and indicates that there are two somewhat distinct subpopulations. This often seems to be the case in upper-level maths courses at university. Half of the students do very well and there's a peak in results at the high end, but half of the students struggle and there's another peak at the low end, with nothing much in the middle. It would also be what happens if we measured testosterone levels across the whole population: there would be a clustering and a peak lower down showing women and a clustering and a peak higher up showing men.

But often, when comparing men and women, the data is not completely separated out like that – there may well be two populations with different peaks, but with so much overlap that just talking about different averages is drastically misleading.

How much overlap there is depends on how 'wide' the bell curve of the normal distribution is. These two pictures show pairs of bell curves where the peak is the same distance apart in each case, but there are very different amounts of overlap:

So just knowing the different means of the two groups is not enough for us to know how different the two groups are overall – that depends on how far apart the means are relative to how wide the bell curves are.

One way of measuring the 'width' of the curve is the standard deviation, so named as it's a measure of how much the data points 'deviate' from the mean. No matter how wide the bell curve is, there is always the same percentage of it within one standard deviation of the mean (on either side). That percentage is around 68. If the bell curve is more tall and thin, this is encapsulated by the standard deviation being smaller, and 68% of the results are then in a much tighter width around the mean. In the following picture the dotted lines show the values that are 'one standard deviation' away from the mean on each side; the distance between the dotted lines is much smaller in the thin bell curve on the right:

In the two pictures of overlapping bell curves shown earlier, the means differed by the same *absolute* amount, but by very different amounts relative to the widths of the curve. In *Inferior*, Angela Saini writes of a table 'more than three pages long' of all the statistical gaps that have been found between men and women on all sorts of measures, including mathematics, aggression and self-esteem. She sums it up like this: 'In every case, except for throwing distance and vertical jumping, females are less than one standard deviation apart from males.' She goes on: 'On many measures, they are less than a tenth of a standard deviation apart, which is indistinguishable in everyday life.' Here are pictures of two normal distributions that are one standard deviation apart and one-tenth of a standard deviation apart, showing that in the former case there is a already a substantial overlap and in the latter case they are essentially indistinguishable:

In *Testosterone Rex*, Cordelia Fine also goes into this, pointing out that there is not a 'sharp line' between men and women, but a 'shifting mosaic of features' rather than a male brain and a female brain. One meta-analysis of many studies found that 'about 40 percent of the time, *at least*, if you chose a woman and a man at random, the woman's score would be more "masculine" than the man's, or vice versa. (If there were *no* average sex differences, this would happen 50 percent of the time, so 40 percent is not far off that parity.)' Again, this included research on a wide range of measures such as mathematical ability, reading, competitiveness and leadership style.

The mistaken idea that men and women are measurably different leads us to regard exceptions to this 'rule' with anything from criticism to outright censure. When we try to sum up gender differences too succinctly, in effect we are making sweeping statements and casting some people in the role of 'exceptions'. The exceptions often turn out to be the result of our definitions or our assumptions rather than anything meaningful. If a female mathematician is considered an anomaly, does that tell us something about women, about mathematicians, or about our preconceived expectations? I would argue it's the latter. It's true that there are many fewer female mathematicians than male ones, but perhaps it is this statistic that should be considered the anomaly. This is all too often to do with what assumptions we are taking as our baseline or default.

The null hypothesis

One objection I often meet when I speak on this subject is the claim that we shouldn't be aiming for equality until we've proved *scientifically* that there are no differences at all. I think this is a question of null and alternative hypotheses.

In a scientific experiment, unlike in a mathematical proof, you have to start by deciding what is the default assumption in the absence of evidence. That is the null hypothesis. Then you look for evidence of something else going on. That is the alternate hypothesis.

For example, in a drug trial, the null hypothesis might be that the drug doesn't do anything besides the placebo effect. The alternate hypothesis is that the drug does more than the placebo effect. If you don't find evidence of anything other than the placebo effect then you revert to the null hypothesis until you can find evidence to show otherwise.

In the case of gender equality, what should be the null hypothesis? When disagreements on gender bias and equality persist even among people who are not overtly prejudiced, they often seem to arise from a disagreement on that point. On the one hand we could take the null hypothesis to be that there is no unfairness, that we should default to assuming that the numerical domination of men in certain fields is because of biology, not bias, until we find evidence otherwise. The anecdotal evidence of women's experiences of prejudice is not counted as evidence as it is anecdotal and not a large, randomised peer-reviewed study. And note that sometimes this is a way to hide overt prejudice in something that sounds justifiable and rational.

On the other hand, we could take the null hypothesis to be that there *is* unfairness, that we should default to assuming that the numerical domination of men in certain fields is because of bias, not biology, until we find evidence otherwise.

Who benefits from these arguments? Men benefit from the first one and women benefit from the second one.

Science is supposed to be impartial, but the choice of null hypothesis in the first place puts an immediate bias into any of these studies. Unfortunately, this is one way in which science has historically been used to hold women back.

And even if we did find that nature played a big part in gender differences in behaviour, it's never going to be deterministic or measurable enough for us to really be able to make predictions based on it. That, together with the fact that it benefits anti-egalitarian people, convinces me that the argument is mainly a distraction and a detraction from the more pressing issues.

We have seen that even if men and women are different there is a large range of behaviour and large overlaps in the range of men's and women's behaviours. I think it would be more reasonable to ask a more nuanced question: instead of adopting a black-and-white position that men and women are different and thus gender imbalance is fair, we could ask about the range of grey in the area: to what extent men and women are different, and how much gender imbalance is thus fair.

However, we have seen the immense problems in the question of measurement. So even if men and women were innately different, the degree of difference might not account for the degree of gender imbalance we currently see; moreover, it is likely to be more or less impossible ever to measure any exact degree of innate difference between men and women, so we will never be able to declare exactly what level of imbalance is 'fair' as an outcome.

We could attempt to control experiments further and further in order to try and get closer to an 'accurate' study, but in a sense this would actually take us further and further from real life.

Life is not a controlled experiment

An even more nuanced question we can ask involves acknowledging that life is in fact not a controlled experiment. One group might be worse than another 'on average' but have the potential to be better if nurtured and supported differently. One example is students who lack self-confidence. The obvious 'received wisdom' is that self-confidence helps you do better. However, students' lack of belief in themselves can lead to them being much better at recognising their weaknesses and improving themselves, as well as being cautious enough to check their work thoroughly and back everything up with evidence and strong arguments; they just need more encouragement and support to get there. Whereas students who are very confident might be better at persevering in an unsupportive environment, but they are always in danger of shoddy work and baseless claims owing to their confidence. On average, all things being equally unsupportive, I would expect the self-confident students to do better. But in a *supportive* environment I would expect the self-doubting students to do better.

This effect is studied in *The Orchid and the Dandelion* by W. Thomas Boyce. The idea is that 'dandelion' children are resilient and are largely unaffected by their environment, and that 4 in 5 children are like this. He calls the other children 'orchids', and argues that, like the actual flowers, they are more sensitive to the environment and more likely to struggle in an unsupportive environment, but have the potential to do even better than the dandelions when nurtured appropriately.

Whether this is true or not, it exemplifies the idea that averages ignore the added nuance of asking in what circumstances different things happen. According to Boyce's theory, dandelions would do better than orchids 'on average' if situations 'on average' are not suitably nurturing (which they probably aren't at the moment). One response would thus be to favour dandelions; a different one would be to provide nurturing environments so that orchids can also reach their full potential to participate in society. Of course, dandelions benefit from not doing the latter, so self-interested dandelions are likely to oppose it.

This leads me to want to investigate whether men are now

more successful than women at certain things just because of the environment that we have set up. And this might be due to things explicitly to do with gender such as prejudice, or things that are only implicitly to do with gender via correlations and observed statistical differences, such as different character traits that are currently more prevalent in men and some that are currently more prevalent in women. I think this means that rather than look for gender differences we should look for character traits that cause different people to respond differently to different situations.

Different answers in different situations

Maths is not just about getting 'the right answers', because different answers can be right in different mathematical worlds. At a basic level, say in school maths, someone else tells you which mathematical world you're required to work in, and then there might well be fixed 'right answers'. In the world of ordinary numbers 1 + 1 is always 2, you can't take square roots of negative numbers, and infinity isn't a number.

But in higher-level maths it becomes more about dreaming up different worlds in which different things can be true, so instead of asking, 'Is this true or false?' we might ask, 'In what worlds could this be true and in what worlds is this false?' At this higher level of maths you can find worlds in which 1 + 1 is 0, or 1, or 3. You can take the square roots of negative numbers if you move into the world of 'complex numbers' instead of ordinary real numbers. Infinity is a number if you move into the world of the 'extended real numbers'.

Our current human world has been handed to us by past generations. In this world men and women might well appear to be different in various ways. Men are more successful than women, overall, in maths and science and business and politics. But the mathematician in me says that there could be a different type of world in which that doesn't have to be true, and not just by us imposing quotas and insisting on fifty-fifty gender ratios, both of which are divisive policies and miss an important point: if something about the environment is stopping women being as successful as men, then imposing these ratios without thinking about the environment is going to increase women's representation without necessarily increasing their success. This is perhaps encapsulated as the difference between diversity (which is about numbers) and inclusion (which is about environment). But that still carries the danger of being stuck talking about gender differences, and thus being divisive and one-dimensional.

In the next chapter I'll look at the flaws in the sort of action we take in response to our ideas about gender differences, and then I will show that there is another approach we can take instead, by dreaming up a whole different world on a new dimension.

3

The problem with leaning in

Here is the kind of question that perplexes children in maths lessons at school:

> *If Alex has 7 cookies and Sam has 3 cookies, how many cookies do we need to give Sam to make sure they have the same number of cookies?*

Well, we could give 4 more cookies to Sam, or we could take 4 cookies from Alex, or we could make Alex give 2 cookies to Sam. In any of these situations Alex might be upset, depending on what sort of person they are. And in the maths class the children might well wonder why we want to do this at all. Children are apt to ask questions that probe the validity of contrived maths problems. What if Sam doesn't even like cookies and would rather have apples?

In higher-level maths, if we're comparing x and y, then instead of asking how to add something to x to make it equal to y we're more likely to investigate senses in which x and y are the same or different, and think about contexts in which things can seem the same or different. Three cookies and three apples are the same in terms of number, but not in terms of nutritional content. Depending on your tastes, they may not be the same in terms of pleasure either.

This requires us to ask deeper questions about what roles *x* and *y* are playing, and what aspects of *x* and *y* we're even interested in in the first place. These are much more subtle questions that require flexibility of thinking, and often the ability to change perspective and find different abstractions focusing on different points of view. Higher-level maths is not about fixed answers, but about flexible thinking and deep questioning.

The issue of how to make things equal is relevant to gender differences. We have become so fixated on thinking about gender differences that we associate character traits with genders and then try to make genders equal by compensating for the perceived shortfalls on one side, like giving Sam more cookies. This is typically flawed twice over: first, by us associating character with gender in the first place, and second, by us assuming that the character traits associated with men are more valuable, and that to make men and women equal we thus need to get women to have more of the character traits associated with men.

I am going to argue that a better approach is first to decouple character from gender, and then to think more deeply about the roles that different character traits play. Some aspects of our actual human experience are inextricably linked to gender, when it comes to discrimination, explicit and implicit bias, structural power imbalances, and many instances of sexual harassment. But character types need not be associated with gender – in principle a man can perfectly well have some character trait that a woman has, and vice versa. If we decouple character from gender,

I believe we can think more clearly about the ways in which we currently overvalue certain character traits and how we can change our assumptions, but crucially also our structures in society, to value character types that we truly think are valuable.

This is hard work, and a big project.

Gendered words

The first part of the work is to stop automatically associating character with gender. At its most obvious, the assumed connection happens via the very words 'masculine' and 'feminine'. What are these words doing for us? They are prescriptive rather than descriptive, in the following sense.

If you hear somebody else (somebody more traditional, perhaps) describe a woman as 'very feminine', or a man as 'very masculine' or 'manly', an image probably comes to mind, even if you wouldn't use those words yourself.[1] But why? The words do not just describe the behaviour of women and men – they prescribe a supposed 'ideal' or 'natural' behaviour of women and men.

They might genuinely describe trends in men's and women's behaviour to some extent, and at some point in the past they might have done so even more. But across history in most (but not all) cultures men and women have been expected to behave in certain ways, and we can hardly separate out those expectations from the behaviour that then resulted. So to say that these gendered words really were

---

1 For an entire book about assumptions around these words, see *Femininity* by Susan Brownmiller.

descriptive of men and women's behaviour in the past is to ignore this inextricable circular relationship between expectations and behaviour:

As we have seen in the previous chapter, the idea of looking for gender differences is at best a distraction and at worst a tool of oppression. When it comes to 'masculinity' and 'femininity', both men and women suffer from pressures to behave in a certain way. If a man is 'unmasculine' or 'effeminate' it sounds like something is not quite right, and similarly if a woman is 'unladylike' or 'masculine'. We have fabricated ourselves a contrived system of judgements like this:

	men	women
'feminine behaviour'	anomalous (bad)	appropriate (good)
'masculine behaviour'	appropriate (good)	anomalous (bad)

We can free ourselves by thinking in terms of independent variables.

Independent variables

In maths, quantities may or may not have a definable relationship between them. When there is a fixed relationship it helps us reduce the number of things in the situation that we need to understand. This is often the point of an equation, such as this one for circles:

$$\text{circumference} = 2\pi \times \text{radius}$$

This is a definable relationship between the radius and the circumference, no matter what size the circle is, and it tells us that we don't need to understand those two concepts separately in order to understand a given circle: we can deduce one from the other.

By contrast, if we think about a cone with a circular base,

the radius of the base and the height of the cone are independent: we can't deduce anything about one from the other. There is no definable relationship between them. To give a less abstract example, in some jobs the salary scale is determined by how long you've been there, in which case you can work out how much someone is paid if you know how long they've been in the job. For example, if you have a fixed-percentage pay rise each year the graph of people's salaries might look like this, with the grey line showing that the situation is a one-dimensional line:

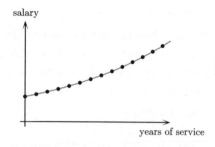

But in many jobs the salary depends more on negotiating pay rises, and so it can be wildly inaccurate to assume that people with the same experience are paid similarly (and, by the by, neglects the fact that men on average apparently ask for, and are given, bigger pay rises than everyone else). The graph of people's salaries might look more like the one below – there is a vaguely upward trend, but any attempt to model it by a one-dimensional line (as shown) is going to be desperately approximate:

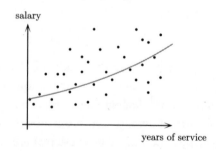

This tells us that, rather than express salary *in terms of* years of service, we would do better to study those variables separately, instead of trying to use one to understand the other.

In more complex situations, an intellectual sleight of hand

(sleight of mind?) happens in several small steps, making each one harder to detect but compounding into a large mis-representation, as with the small slides compounding into weak arguments in the previous chapter. One contentious example is body mass index, or BMI. The formula for BMI uses height and weight, so the relationship between height, weight and BMI is defined and fixed. However, BMI is then taken as a proxy for body fat, which is in turn taken as a proxy for healthiness, even though those relationships are not fixed or precisely definable. So we have the following chain of variables that we are considering:

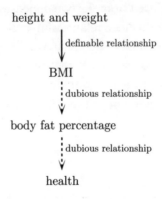

When we take one quantity as a proxy for another we are implicitly assuming that the two have a fixed relationship. When that isn't true, as in the case of BMI and health, or gender and character, we risk oversimplifying or even misinterpreting a situation. We can escape this by acknowl-edging that they are independent variables and considering the two separately. This means that we have moved into a higher dimension, because we now have to understand two things instead of one. For example, when we draw basic graphs in maths we often study a relationship between x and y such as:

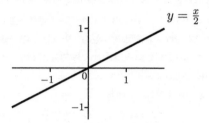

The graph of this relationship is a line, that is, it is one-dimensional, reduced from the two dimensions of the whole plane because of the fixed relationship between x and y. We have lost a dimension. Whereas if we acknowledge that x and y are independent, we get that higher dimension back – there is nothing to reduce our situation from two dimensions, and so we could be anywhere in the two-dimensional plane:

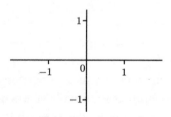

Going into higher dimensions makes things harder, which is one reason why we might be reluctant to do it. But I think this is what we need to do with gender and character: acknowledge that using one as a proxy for the other is flawed, and that we need to treat the two as independent variables. When gender is relevant, we ought to consider it.

When character is relevant, we need to consider it. But we shouldn't assume they are linked. In the rest of this chapter we will examine the consequences of *not* using the higher dimension, and in the second part of the book we will argue that the payoff for going into the new dimension is large and the effort involved comparatively small, but we are currently not daring to go there.

The current one-dimensional approach

Currently, instead of considering gender and character separately, we are doing various confused things. On the one hand, in the name of equality women are encouraged to be more like men in order to be 'successful', and men are encouraged to get in touch with their 'feminine' side, to learn empathy, express emotions and be collaborative, although these are not gender-specific traits: people of any gender can have empathy and emotions, it's just that certain traits have traditionally been associated with certain genders, and we're beginning to see that that isn't necessary. On the other hand, there's a reaction to this that is sometimes referred to as 'toxic masculinity', a response in which men, perhaps feeling disenfranchised by women encroaching on their roles, revert to an extreme version of traditional masculinity involving violence, often sexual violence in particular, and arguing for the reversion to male power roles and female submission. Here is a diagram depicting those tensions:

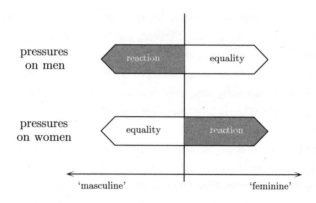

It is worth noting that if women become more like men, then certain men feel threatened. Unfortunately I have long experience of this since I am successful in mathematics, a field that is traditionally associated with men. I have seen men feel threatened by me in social situations, in professional situations, and in random interactions online. I once went to a party and tried introducing myself as working in a more traditionally female field (human resources) and had a much more friendly response from men. It's true that this is just my personal experience, but if you pick any successful woman scientist on social media I'm certain you'll see it happening.

In fact this phenomenon has been studied in controlled experiments, with articles such as 'A man's (precarious) place: Men's experienced threat and self-assertive reactions to female superiors'.[2] Another study[3] found that men felt

---

2 Ekaterina Netchaeva, Maryam Kouchaki and Leah D. Sheppard, *Personality and Social Psychology Bulletin*, vol. 41, no. 9 (2015), pp. 1247–59.
3 Lora E. Park, Ariana F. Young and Paul W. Eastwick, '(Psychological) distance makes the heart grow fonder: Effects of

attracted to women more intelligent than them as an idea, but were attracted to less intelligent women when it came to actually meeting them. Unfortunately, I've experienced that rather a lot in the past too.

On the other hand, when men become more 'like women' in the sense of being empathetic, communicative, in touch with emotions, then I feel supported not threatened, and I'm sure it's not just me; however, I've seen other men feel it as a threat to the entire concept of masculinity, claiming it's a 'war on men' or an 'attack on masculinity'. It is not a war on men; it's a war on the pressures of traditional gender roles, and in fact it's not even a war, more of a gradual attempt to change culture.

A widespread response is that we need to redefine masculinity to give men a more clearly explained role in the new age of gender equality, or to persuade men that masculinity includes compassion and caring and respect for women. Brad Pitt has been talking about his recent film *Ad Astra* in terms of rethinking the definition of masculinity, although according to Steve Rose the film is another in a long history of gender-stereotypical space films involving 'heroic men and emotional women'.[4]

There is a backlash from women as well. Some women say they like being 'very feminine'. But women can still be whatever it is that they currently think of as 'feminine' – it just doesn't need to be associated specifically with being a

---

psychological distance and relative intelligence on men's attraction to women', *Personality and Social Psychology Bulletin*, vol. 41, no. 11 (2015), pp. 1459–73.
4 Steve Rose. 'Fly men to the moon: Ad Astra and the toxic masculinity of space films', *Guardian*, 16 September 2019.

woman. Unless it involves something like biologically giving birth to a child,[5] men can do it too, whether it's wearing slinky dresses, styling your hair, making an art of make-up, or having an alluring way with men. And if another woman doesn't do those things, then that shouldn't be called 'unfeminine'. Neither person should, in my view, have more or less claim to being female or woman-like; that idea keeps us stuck in one dimension, and, worse, a skewed single dimension.

The skewing of the system

Being attached to a particular dimension or point of view isn't necessarily a bad thing, but in this case the gender dimension comes with embedded bias: inequality gets built into the system by a tendency to overvalue traits tradition-ally associated with men and undervalue ones traditionally associated with women. Take the debunked image of hunter-gatherers: we now know that around two-thirds of the food came from gathering and only one-third from hunting.[6] The hunt was dangerous and often unsuccessful, but – surprise – is still popularly celebrated and used to prop up justifica-tions of male domination.

Fast-forward ten thousand years or so to Bletchley Park in the Second World War, the site of the famous codebreak-ing that would eventually help defeat the Nazis. The work was spearheaded by Alan Turing and other mathematicians,

---

5 Some transgender men might be able to give birth to a child as well.
6 See Saini, *Inferior*, and *The Creation of Patriarchy* by Gerda Lerner.

but required the collaboration of thousands of workers, around three-quarters of whom were women. Many of them performed routine clerical tasks, but others held degrees in mathematics, physics and engineering and did specialist codebreaking work. Still, all of the work was crucial, including operating the codebreaking machinery, transcribing the coded messages in the first place, determining when a decoded message was recognisably German, and translating it. Of course, the leading mathematicians were mostly male because of how the education system excluded women; the rest of the workers were mostly female as men who weren't expert enough for the leading maths roles were away fighting.

It may well be that Alan Turing deserves the greatest credit because of the unique and unreplicable role he played in breaking the codes, but until quite recently a rather black-and-white situation persisted in which the men had all the credit and the women had none. Even when women's roles have been acknowledged it has typically been as 'support' workers to the men's roles of 'genius'. A lone eccentric male genius is of more interest than thousands of hardworking women. In *The Bletchley Girls*, Tessa Dunlop seeks to fill in some of the grey area by telling the women's stories, from their point of view.

In the end the achievements of Turing were also undervalued: he was gay and treated abominably by the British authorities instead of being hailed as a hero. Being male, it seems, is necessary but not sufficient to be considered a true 'hero'; it is not only a patriarchy we are dealing with, and a white-ruled one, but a heteronormative one with many other oppressive qualities besides.

The leading mathematicians at Bletchley were selected

through a drive to recruit 'men of the professor type', especially from the universities of Oxford and Cambridge. It is not clear what a 'woman of the professor type' would have meant to anyone at the time; Dorothy Garrod only became the first woman Oxbridge professor in 1939. As higher education expanded (finally) to allow women in, the image of a genius as necessarily male has perhaps weakened, but has it weakened enough? Recent data shows that, while far more women have careers in academia now than seventy years ago, the situation is still very unbalanced: there are many more men than women at more senior levels, and more women than men in part-time positions, which are automatically untenured and thus less respected in many places, including the US.

But women may well be less respected regardless of their job title and achievements. Historically, high-achieving women are either less famous or remembered in selective, equivocal, or downright insulting ways. The 2013 *New York Times* obituary of the scientist Yvonne Brill began like this:

> She made a mean beef stroganoff, followed her husband from job to job and took eight years off from work to raise three children. 'The world's best mom,' her son Matthew said.[7]

Only in the second paragraph did the obituary note, 'But [she] was also a brilliant rocket scientist...'. Not only did

---

7 Margaret Sullivan, 'Gender questions arise in obituary of rocket scientist and her beef stroganoff', The Public Editor's Journal, *New York Times*, 1 April 2013.

it lead with domestic descriptions, it then seemed to assert the improbability of being a good mother and a rocket scientist, inviting us to be suprised that anyone might be both. The *Times* quietly edited it after a general outcry, but only replaced the opening beef stroganoff reference with a reference to rocket science, and left the rest of the domestic introduction as it was. In fact, Yvonne Brill devised and patented a propulsion system for satellites that is still used today. The *New York Times* now runs a series of 'Missed Obituaries' of people whose achievements were neglected in the past, typically because they were female or non-white.

A related example is Florence Nightingale, known as the 'Lady of the Lamp' and arguably most widely remembered as a nurse. In fact she was also a ground-breaking mathematician whose nursing was made effective by her analytical approach to data. Crucially, her innovative visualisations of that data meant that the analysis was actually taken on board by those in power, enabling her to make the sanitary changes that would dramatically improve on the dire mortality rates in hospitals during the Crimean War. But 'Lady of the Pie Charts' doesn't sound as romantically evocative as 'Lady of the Lamp', and besides, she doesn't fit the popular image of a scientist: an old (white) man with mad hair, in fact, the exact image of Albert Einstein. Indeed, Einstein is much more famous than the mathematician who gave him crucial help in his work and who happened to be a woman, Emmy Noether.

Emmy Noether

Emmy Noether is very famous among mathematicians and

physicists, but almost unheard of outside mathematics. (Perhaps that isn't surprising given how few mathematicians are heard of outside mathematics.) However, given the rise in literature about and general attention being given to previously 'unsung heroes' who were women, her prominence is rising and she usually features on lists of brilliant women mathematicians, scientists, or 'geniuses' we should have heard of.

Emmy Noether was at the forefront of women in mathematics, in that she was in the first wave of women who were allowed to reach certain levels at each stage of her career. She was born in Germany in 1882 and attended Erlangen University at a time when women were only allowed to observe, not fully enroll as students. That had changed by the time she did her PhD, and she was then the first full-time female student at Erlangen. She went on to teach there – without pay, as women were not yet allowed university positions.

Her work came to the attention of some eminent mathematicians in Göttingen: Felix Klein (of Klein bottle fame) and David Hilbert (who has Hilbert spaces named after him), who were working on Einstein's new theory of relativity, and exchanging letters with him about it; those letters are very helpful for historians, and give us insight that we can't get when collaborations happen face to face.

They were stuck on a question about conservation of energy and they knew they didn't have the mathematical understanding to sort it out, so they called in Emmy Noether for help. Letters between Hilbert and Einstein indicate that her contribution was crucial and that they both respected and deferred to her expertise. At one point Hilbert includes a note from Noether to avoid a long explanation, and at another Einstein writes: 'How can this be clarified? Of course it would be sufficient if you asked Miss Noether to clarify this for me.'[8]

Meanwhile, Noether was not allowed to have a teaching position because she was a woman, but for a while she gave lectures under the name of Professor Hilbert. The lectures were advertised as being given by Hilbert 'with the assistance of' Noether, and everyone in the know understood that to mean that she was actually giving the lectures.

Einstein and Hilbert both petitioned for her to be given a teaching position, arguing that it was irrelevant that she was a woman because, as Hilbert famously pointed out, 'This is not a bath house.' There was resistance from other faculties but she was eventually granted the lowest form of teaching position, not a real professorship, but still at least she was then paid (a little) instead of teaching for free.

This story unfortunately does not have a happy ending. As well as being a woman, Noether was also Jewish, and soon the Nazis came to power and banned Jewish people from civil service positions. University posts counted as civil service and so in 1933 Noether's position was withdrawn. She came as a refugee to the US where she was given a post at Bryn Mawr College, but there is still no happy ending: in 1935 she was operated on for a large ovarian cyst, and although the operation appeared to be successful she developed a fever and died a few days later at the age of just 53.

So much for Noether's life and travails. What about her work? Her great contribution to physics and to Einstein's theory of relativity in particular was work connecting

---

8 See Yvette Kosmann-Schwarzbach, *The Noether Theorems*.

mathematics and physics, specifically connecting symmetry in mathematics with conservation laws in physics. This insight did not directly solve the problem of conservation of energy that Einstein had, but it provided the illumination needed to help solve it. The question was: what happened to the laws of conservation of energy? Noether's work said: conservation laws in physics come from symmetries in the underlying mathematical system, so look for the mathematical symmetries and you will find the conservation laws. Noether's insight enabled a flow of understanding from one field of research to another.

Unfortunately for Noether, those sorts of moments are not as dramatic as the kind where you make a big leap forward, as Einstein's theory of relativity did. Noether was thanked and appreciated in many personal letters sent between Einstein, Klein and Hilbert as they grappled with these issues. Klein wrote to Hilbert: 'You know that Miss Noether advises me continually regarding my work, and that in fact it is only thanks to her that I have understood these questions.' However, the extent to which they depended on Noether's help is less evident from the formal citations in the published papers. This might be technically correct if the theory of relativity did not directly depend on her work in a technical scientific way. Perhaps her work was more about shining light to help others find the way, rather than producing bricks and tools for them to use directly in their building. The people who produce bricks and tools are the ones who are officially credited, rather than those who shine light; is this fair?

Noether's work in pure mathematics also shone great light, and arguably was the beginning of a new sort of abstract mathematics that took off in the later part of the twentieth century; the birth of my field, category theory, was part of the journey towards greater and greater abstraction as a means of unifying mathematics. Category theory is sometimes criticised for building theories rather than proving big theorems, and sometimes it is criticised for 'merely' shining more light on things that have already been proved.

Proving big theorems and solving problems that are hundreds of years old are a bit like the mathematical versions of climbing Mount Everest or going to the South Pole for the first time. There is something non-incremental about it and thus we might call it 'macho', falling yet again into the fallacy of attributing gender to a character trait. Once we attach the idea of maleness to a trait we are in danger of overvaluing it and then asking everyone to emulate it.

Emulating male behaviour

When patriarchal society overvalues character traits associated with men there are broadly two common responses. One is to claim that gender differences are innate, and that women thus can't expect to be the same as men; the other is to urge women to emulate men in order to be successful.

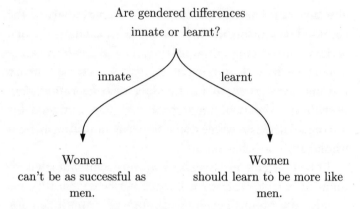

Are gendered differences
innate or learnt?

innate   learnt

Women
can't be as successful as
men.

Women
should learn to be more like
men.

Men may claim that differences are innate in order to exclude women from male-dominated activities, and to relieve themselves of the responsibility of having to do anything about gender imbalances. But women may say it too: successful women, in order to retain their status as a rarity, and others to relieve themselves of the pressure of trying to emulate men or strive for success according to the existing definitions of success. After all, women may not want to emulate traditionally masculine behaviour (and in fact men may not want to either). If the choice seems to be between emulating that behaviour or not being successful, many people may choose the latter. But that choice is a false dichotomy, based on flawed underlying beliefs.

I have my own personal story of emulating the behaviour around me to become successful at something predominantly done by men, that is, mathematics. My story goes like this.

When I started as a PhD student I decided I should hide all aspects of femininity as much as possible, so as not to give anyone a chance to stereotype me and say I was no good because I was a woman. Without being explicitly conscious of it, I started emulating behaviour of successful people around me in order to be successful, and as the environment was male-dominated that meant I was learning to emulate the behaviour of men. But this doesn't quite make sense, as of course there is a whole range of behaviour of men and a whole range of behaviour of women.

In fact, after several years of university life mostly surrounded by male professors, I moved to a women's college as a junior research fellow and met a large number of women professors for the first time. I found that the environment was, disappointingly, not very different, because so many of the women fulfilled the dominant and domineering roles that men might otherwise have occupied. The women were very successfully emulating stereotypically male behaviour, and eventually I found I was doing it too: I was learning to be competitive, ambitious, assertive, unyielding. When I found myself in a high-powered academic environment I discovered that certain people – not all – were prone to be argumentative, sceptical, dismissive, and instantly ready to jump up and challenge anything that anyone said. I tried to learn to do the same.

In *No Contest* Alfie Kohn describes 'pseudofeminism' as seeking 'the liberation of women through the imitation of men'. It is a compounding of two assumptions: first, that gender differences are tied to character differences, and second, that the character types associated with men are more valuable, or even critical, for success. Of course, it's not just women who are imitating men; it's men too.

Emmy Noether, Florence Nightingale, Yvonne Brill and

the women of Bletchley are just a few examples of how women have been successful without necessarily emulating the behaviour of men. Unfortunately, to counter the effect of women being only valued for traditionally acceptably 'feminine' pursuits, there is a tendency then to celebrate them for being the opposite. There is a recent glut of books celebrating women who were 'just like men', with titles emphasising that women can do everything men can do, or emphasising traits of these women that are not the traditionally 'feminine' ones. I fall over them any time I look for books on female role models from history, with titles including words like 'Headstrong', 'Badass', 'Rebel'.

Sheryl Sandberg's book *Lean In* seems to urge women to adapt themselves to the (male-dominated) business world in order to be successful in it. It has been criticised in many ways, one of which is that it seems to assume that in the current structures women can succeed if they just try harder, regardless of built-in sexism, racism, and other inequalities of access. And 'trying harder' all too often seems to consist of trying harder to behave like the successful men. I don't think our aim should be to show that women can be 'just like men'. I don't think we should have to imitate or emulate typical male behaviour in order to be successful. Jessa Crispin includes this in her bracing takedown of some common forms of feminism: 'a fight to allow women to participate equally in the oppression of the powerless and the poor'.[9]

Making sure women can do all the things men do is some sort of progress, but it doesn't necessarily represent or cause

---

9 Jessa Crispin, *Why I Am Not a Feminist.*

a great leap forward for all other women. Sheryl Sandberg's approach has been criticised as a 'trickle-down' version of feminism.[10] Having Margaret Thatcher as prime minister didn't exactly bring about the end of discrimination against women in the UK, and the glass ceiling is alive and well despite her apparent shattering of it.

Asking women to behave like traditionally successful men assumes that that sort of behaviour is crucial for success. It is what I referred to in the previous chapter as the assumption 'Y is important for Z', where Y is some quality that men are perceived as having more than women. Thus, if women must be successful, they must learn quality Y, that is, emulate the men.

What traits are valuable for success?

When I reconsider the behaviours I learnt during my time in mainstream academia, I wonder if some of them were valuable. Perhaps being able to be uncowed by aggressively dominant people is a genuinely useful skill to have learnt, but it's only useful because aggressively dominant people exist. If they didn't exist it wouldn't be a useful skill at all.

In any case I eventually hit a limit as to how much of this behaviour I wanted to learn. Once I was tenured I felt more secure and started allowing some 'femininity' in, but only outside work. This created a strange double life, so I tried to merge the two and be feminine at work too. That's the narrative I gave myself, anyway – I find it interesting to note that

---

10 See, for example, Melissa Gira Grant, '"Like" feminism', *Jacobin Magazine*, 4 March 2013, among others.

this meant something to me at the time, but I've now moved so far from associating character traits with gender that it's hard for me to reconstruct it. I think 'femininity' at the time meant things to me such as showing vulnerability, not demanding to take charge of situations all the time, listening to other people more, and occasionally wearing a dress.

I realised that I didn't like the stereotypically masculine behaviour, and didn't like myself when I behaved like that. I had been learning behaviour I didn't like. But I couldn't work out how to unlearn that behaviour and stay in academia; it felt as if the two were inextricably connected. Then I realised I wanted to take my career in a rather different direction, and over the course of a few years I changed my life around completely. I've built a new career in which I really don't have to encounter those types of people any more, so that I rarely have to invoke those skills. Even if I do encounter such people I try to behave differently now, in order to defuse rather than escalate that type of interaction, and I wonder whether the competitive behaviour I learnt was important or not, even for academia where I learnt it. Was there another way to be successful in that environment? More radically, was there another way for that environment to be?

When I was still in the process of changing my career around I had another experience that reminded me of my time in the women's college and galvanised me further. I met a woman at a party who mentioned her $5 billion budget in her first sentence, along with her name. She proceeded to be very combatively forthright about her success in a male-dominated business world, even going as far as boasting that she could compete with any guy when it came to 'locker room talk'. Her body language was not just strong but aggressive, and she seemed to invite people to challenge her just so that she could take them down a few notches, possibly by comparing the size of their respective budgets.

One might be tempted to call such a woman 'masculine' and she might take it as a compliment. She did, after all, seem to have emulated 'male' behaviour in an expert fashion in order to rise in a competitive business world. I tried to point out to her that I have not had to emulate men's 'locker room' behaviour in order to be successful (we should also acknowledge that most men do not engage in so-called 'locker room' behaviour either). She replied with a sneer, 'Oh yeah, and what makes you think you're successful?'

Although I was very bothered by this encounter at the time, I'm now grateful to the 'Five Billion Dollar Woman' for prompting to me to think really hard about what I do consider to be 'success'. I started out assuming that I wanted to be successful in an already accepted route, by getting a PhD, having my original research published in international journals, winning tenure, getting promotion. Eventually I realised that those definitions of success are a bit meaningless to me. I was expected to apply for grants in order to get promotion. But I realised I didn't really desire promotion and I didn't really need grants either. For me, success is about the people whose lives I have helped. As an educator my biggest opportunity for helping people is in enabling them to understand something they find difficult, for example mathematics. This is a different type of success. It is a type of success that is not celebrated as much by society, and it took some effort for me to realise that my values aren't the same as the one the standard academic career path pushes.

Success in mathematics

One traditional marker of success in mathematics is the Fields Medal, the prestigious prize for mathematics that is sometimes described as the 'Nobel Prize of mathematics' because there is no actual Nobel Prize in mathematics. No woman won the prize until 2014. The first woman to win it was Maryam Mirzakhani and she was also the first Iranian to do so; tragically she died in 2017 of breast cancer. She described herself as a 'slow' mathematician, and worked on problems by doodling on large pieces of paper, which her young daughter described as 'painting'.[11] This is a far cry from the idea of mathematicians as competitive machines, solving problems at speed.

A marker of 'success' at an earlier stage in mathematics is the International Mathematical Olympiad, an international maths competition for school students. The training that some of these youngsters undergo for this competition is arguably as serious as training for the actual Olympics. And it is not that rare for the winning team to be all boys. According to the Five Thirty Eight website, in the history of US participation (which began in 1974) 88% of its six-person teams have been entirely male.[12] The US won in 2018, 2016 and 2015 with all-male teams each time. (In 2017 South Korea won and had one female team member.)

There is ongoing debate about the lack of girls at high levels in this competition. Are they not chosen for the team

---

11 Andrew Myers and Bjorn Carey, 'Maryam Mirzakhani, Stanford mathematician and Fields Medal winner, dies', *Stanford News*, 15 July 2017.
12 Leah Libresco, 'Girls are rare at the International Math Olympiad', FiveThirtyEight.com, 22 July 2015.

because of bias, or are they just not as good at maths? Some people point out that they can't be chosen for the national teams because there are hardly any girls competing in the regional or local teams. Why is that? How can we persuade more girls to take part?

But I would rather pose the question: is it important for us to persuade more girls to take part? The Olympiad is not the be-all-and-end-all of mathematical achievement. It is a constructed competition in a field that really doesn't have competition as its focal point. The Olympiad teams do often involve people who go on to become brilliant or great mathematical researchers, but there are plenty of other ways to become brilliant mathematicians as well. In fact, maths competitions run the risk of actually putting off some people who like maths but don't like competitions: such as me. (Incidentally, Maryam Mirzakhani did attain the highest level at the Olympiad in 1995, achieving a perfect score and two gold medals despite later calling herself a 'slow' mathematician.)

I don't think we necessarily need to try and find ways to get girls to succeed at those competitions; I don't think that's the right aim. I would rather we identify worthwhile strengths that do not shine in competitions, or that are not recognised by the other traditional measures of success, and find ways to encourage and value those strengths, regardless of anyone's gender.

Instead of getting women to imitate or emulate men, we can value them as people in their own right. Much has been written on the strengths that women bring to teams and to management, although as usual we risk making sweeping statements and having to qualify them by pointing out that not all women have those strengths and many men have them too. These are strengths such as communication skills, empathy and collaboration, which are valuable to any group of people trying to achieve something together.

However, valuing women for these 'feminine strengths', and valuing men when they exhibit those traits, does not address our fallacious assumption that character traits are tied to genders. It doesn't free us from the divisive gender debate. It doesn't take us out of one dimension. If we object to the idea that 'men are better' it's not that helpful to declare instead that 'women are better'. It pits men and women against each other and sets up a prescriptive framework rather than a descriptive one.

In maths we only use descriptive theories. There is no way for an abstract theory to impose behaviour on concrete life, and so if our theory doesn't accurately describe something then it might be a logically sound theory but it won't be relevant or helpful. Theories about people can influence people's behaviour, especially if the theories are believed by a significant proportion of society, but that doesn't make them helpful; in fact it makes them more destructive, where the mathematical version would just be irrelevant without really doing any harm.

Instead we can do what maths does when things have become a bit confused: we go back to first principles.

Going back to first principles

In maths, going back to first principles usually means working out what assumptions we are relying on, and shedding as many as we can. It means digging deeper through our thought processes to go back to much more fundamental concepts.

For example, when a woman is sexually assaulted, attention too often turns to what she was wearing. But if someone claims that she was sexually assaulted because she was wearing revealing clothing, she is implicitly assuming that men (as it is usually a man) are unable to prevent themselves assaulting a woman in revealing clothing. This diagram shows the combination of those factors, with the dashed line showing the part that is too often overlooked:

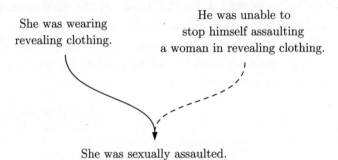

She was wearing revealing clothing.

He was unable to stop himself assaulting a woman in revealing clothing.

She was sexually assaulted.

To pick a less violent example, when I turn up for a speaking event wearing a dress, the tech guy quite often says that it will be hard to mic me because I'm wearing a dress; I typically interject, '... and because the equipment is designed for men' to draw attention to the baseline assumption that this is the only possible way microphones could be designed.[13]

---

13 Note that the tech person is usually a guy but not always, and the non-male tech people I've had have never said this to me. Also, I now carry around my own gear that makes it possible for me to wear lavalier microphones designed for men even when I'm wearing a dress with no belt, no collar and no pockets.

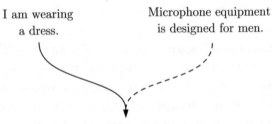

I am wearing
a dress.

Microphone equipment
is designed for men.

It is difficult to attach a
clip-on mic and transmitter to me.

Note that there is another assumption here, that men will wear clothes with a collar and pockets; we could have added in a dashed line connection for this.

Now let us examine the assumptions contributing to the idea that 'character traits traditionally associated with men are better for success':

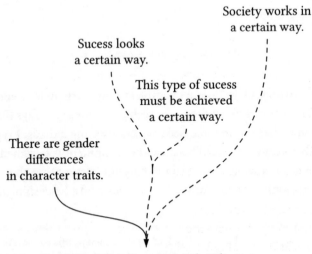

Society works in
a certain way.

Sucess looks
a certain way.

This type of sucess
must be achieved
a certain way.

There are gender
differences
in character traits.

Traits traditionally associated with men
are better for success.

In this case this means shedding the assumption that gender is tied to character and separating out those variables. This means focusing on gender when that matters, and focusing on character traits otherwise.

It's important to remember that in many senses gender does affect people's experiences, and experiences affect behaviour. I am not saying we should be 'gender-blind'. While inequities continue to exist in the system, and while historical inequality still has an impact on the present, I advocate taking much more active steps to counteract this. But the steps we need to take can, I think, be understood more clearly if we consider gender and character separately. The story of Dame Stephanie Shirley is a case in point.

Dame Stephanie Shirley

Dame Stephanie Shirley was at one point the only woman on the *Sunday Times Rich List*, so even the Five Billion Dollar Woman might deign to call her successful. She was a pioneer twice over – first in computer programming (which is how she made her money) and then in the care of autistic children (which is how she spent it).

I learnt her story from her own memoir, *Let It Go*. She and her sister arrived in London on the *Kindertransport* in 1939 when she was five, and they were cared for by foster parents in the Midlands. She had a flair for mathematics and encountered gender-based obstacles early: her girls' school did not teach mathematics and so she had to go for maths lessons at the local boys' school.

She became a computer programmer in the fifties, one of a very small number of them who happened to be women.

She found it extremely difficult to be taken seriously or indeed hired, especially after having children, and she also writes of experiencing sexual harassment at work. So she decided to set up and run her own company from home instead, and specifically tap into the pool of neglected talent that women computer programmers represented.

At first she had some of the same problems – she couldn't get potential clients to take her seriously. So then she tried an experiment – she sent out the same letters seeking business, but instead of signing them 'Stephanie' she signed them 'Steve'. Work started coming in.

While this might sound like emulating male behaviour, it's really just using a male name in order to overcome bias, not unlike Charlotte Brontë publishing *Jane Eyre* under the pen name 'Currer Bell', Mary Ann Evans writing as George Eliot, or Joanne Rowling as J. K. Rowling.

Stephanie Shirley specifically did not emulate male behaviour in the running of her company. In fact, she more or less did the opposite – she looked at the reasons women were being excluded from the industry, saw them as strengths, and built a phenomenally successful company on that. What were those reasons?

First of all there's all-out prejudice. This meant that Shirley believed there were women programmers out there who were better than some of the male ones but weren't being hired, just because they were women, and especially because it was standard at the time for women to stop working when they got married or when they had children. Thus, hiring them would already put her company in a strong position. But then there was the more tangible issue of women having babies and then only wanting to work part-time, and perhaps only being able to work from home with irregular hours. Not hiring women on those grounds is now seen as something that contributes to overall structural sexism in society, but many people including traditionalists, more hard-nosed capitalists, and people who run small businesses still object to this idea, because they think that anyone who wants to work short, irregular hours from home is genuinely a liability, whether they're a man or a woman with children or not.

This has to some extent been addressed by legislation forcing companies to disgregard this fact, in the sense of allowing women to take maternity leave (and to some extent permitting men time off as well), and allowing them greater flexibility when their children are very young. However, it is worth noting that companies still find ways to get round this, and in the US the rules are very weak compared with other developed countries.

Stephanie Shirley took this further and, rather than just disregarding the issue, she built her company on it, specifically hiring mothers of young children. Writing code is, after all, something that even in those days did not need to be done in a specific location at specific hours (programs were written on paper and only implemented on a computer later). Shirley's view was that mothers with small children had the potential to be the most motivated and efficient of all workers, knowing that their uninterrupted time was limited, but desiring, as women's liberation spread, to have more intellectual stimulation and to contribute to their family's finances. Importantly, she paid for the work done rather than paying by the hour, thus valuing efficiency over presenteeism, and perhaps playing to the strengths of mothers who knew they had limited time while their children were napping and before their husbands came home expecting dinner to be on the table.

Stephanie Shirley's story has a somewhat annoying sequel, which is that when the Sex Discrimination Act of 1975 came into force her company was one of the first to fall foul of it, and she was no longer allowed to favour women in her hiring. It is a sad indictment that a law brought in to stop the marginalisation of women was immediately used against a company that had been champions of women from the start. However, we can learn from all of her story, including this part.

The story operates on two dimensions simultaneously. There is the dimension that is about explicit and implicit discrimination against women, including girls' schools not teaching maths, women not being expected to go to university, workplace sexual harassment, and women being expected to stop working when they got married, or certainly when they had children. There is the dimension that is about cultural and structural issues that made it difficult for women to thrive, especially after having children, including the culture of presenteeism and inflexible working hours.

Shirley addressed both of these issues by hiring only women, especially those with young children, and by creating a different sort of working environment in which they could thrive. In modern-day parlance, hiring more under-represented people is about diversity, whereas creating an environment in which they feel included and can thrive is about inclusivity. In Dame Stephanie's story, and in general, the two are related, but as it's not a fixed or clearly definable relationship I think it is illuminating to consider them separately to make sure we are not hiding assumptions or missing issues.

To address explicit and implicit bias we may need explicit interventions. However if we *only* address that type of bias then we end up with more diversity and not more inclusivity. If we hire more women but maintain a work environment in which they're miserable then they'll either quit, or stay around miserable and thus unable to work well, which is in turn likely to perpetuate the idea that men are better at the work. There is also the danger of being accused of 'sexism against men'.

If, by constrast, you only address cultural issues then you may be subscribing to the idea of gender blindness, or the idea that we should 'just' treat everyone as people and value them for what they bring, regardless of gender. This would be all very well in an already egalitarian world, but we're not there yet, and this approach fails to address past or broad inequalities in society.

We really need to address both as in this chart:

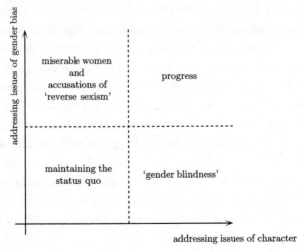

Dame Stephanie Shirley did address both. As a result her company was extremely successful, even by traditional capitalist measures.

At the time, an extreme intervention to overcome previous gender discrimination was arguably required. Maybe we no longer need interventions as extreme as entire companies only hiring women, but we must still address both issues of gender and environment. I think it will help to have a way to think about character and culture as separate from gender, and use this to consider how to change our structures as Dame Stephanie Shirley did.

Changing the culture

Instead of giving women extra help, some of these things could be addressed by changing the entire culture of what is valued and what isn't valued. We could recognise that some skills traditionally associated with women are very important for any community, institution or organisation, such as nurturing, communication, getting the best out of people, and that these skills may go along with having sensitive emotions. Universities could value teaching, mentoring, outreach and public engagement more, and instead of only valuing research in terms of specific results they could also value the organising of conferences, which is often the important catalyst for the research to get done at all.

There was an investigation in the nineties into why male undergraduates did so much better than female ones at Oxford and Cambridge, even though the exams were anonymous and so it couldn't really be put down to direct bias.[14] Perhaps surprisingly, the gender difference in history exams was even bigger than the one in maths.

The finding that stuck with me was that men tended to write essays that took a strong position and argued it fiercely, and that this was highly valued. A balanced position argued from all points of view was valued less. (The report had many more facets than this.) One solution is to train women to make more one-sided arguments. But is that really beneficial to society? We could instead actually think about the value of balanced arguments, regardless of the gender of the person making them.

Rather than celebrating Emmy Noether for being a 'brilliant woman mathematician', we could value collaborative contributions that are the foundations of progress and not just the grand results of figureheads. Another example is Rosalind Franklin, whose contribution to Crick and Watson's discovery of the structure of DNA was arguably undervalued until quite recently. Like Noether, she died tragically young and, like Noether, her work and insights were crucial to a big breakthrough. But she didn't make the big breakthrough. This is not to say that the foundational work is as critical as the big breakthrough idea, but perhaps it deserves more recognition than it has historically received. And, as it happens, it's often done by women.

What would the world be like if everyone collaborated instead of competed? I will come back to this question later,

14 N. G. McCrum, 'The academic gender deficit at Oxford and Cambridge', *Oxford Review of Education*, vol. 20, no. 1 (1994), pp. 3–26.

but I'd like to give a more trivial but still indicative example. I think of it every time I'm waiting for my checked luggage to appear on the carousel after a flight. When everyone crowds around right next to the conveyor belt it is extremely difficult to see what is going on. Nobody can see their luggage until it's right in front of them, and then there's a mad scramble to grab it in time as it glides past.

Whereas if everyone stood back a few steps then we'd all have a better view of our luggage as it approached, and we could then step forward in time to retrieve it in a less panicked fashion, plus there would also be space for us to haul it off without bashing into people. However, if everyone stands back then inevitably someone will take a little step forward to get a better view than everyone else, and then someone else will step forward to see past that person, and so on, and pretty soon everyone is crowding around again and nobody can see.

There would be an uproar if I claimed that stepping forward and getting in everyone's way is typical masculine behaviour, and that considering the experience of other people is feminine behaviour. There would be an uproar if I even claimed that whenever I experience it I see that it is men who are the first to crowd around and get in everyone's way; most people would see through my claim of it being 'my experience' and consider that I was just being sexist. A divisive argument about stereotypes would then follow, and the principle of a better system of luggage retrieval would be forgotten.

This is why we need different vocabulary to talk about character traits. We need vocabulary that is not associated with genders, to give us a way of avoiding those futile and divisive arguments. But we don't just need new vocabulary, we need a whole new way of thinking about and structuring the world. A whole new theory of people.

Part II

UNGENDERED THINKING

4

A new dimension

During the autumn of 2016, the Chicago Cubs were making their way towards winning that thing they won. All sport confuses me and American sport confuses me most of all, but I believe the Cubs play baseball and that the thing they won was a big deal and something they hadn't won for a long time.

During the build-up it seemed to me that the whole of Chicago was going crazy about this except me. This included people who don't usually pay that much attention to sport, such as musician friends of mine and many women.

On the day that they actually played (and won) the last phase (game? match? tournament?) I was scheduled to give a talk for the Illinois Humanities Project. It was in a series of events for practising artists, consisting of a talk, not necessarily about art, and then dinner by a local chef, during which discussion could continue.

I started wondering if I should suggest rescheduling. When we booked it we had no idea the Cubs were going to get this far, and I was worried that in the event nobody would turn up. But people did turn up – a large number of people, in fact – and then I realised that it was a fascinating filter: it was a bunch of artists who really didn't care about a sports team winning something. My students, who at that point were all art students, felt similarly. We all shrugged together and were a bit nonplussed by the levels of excitement.

It got me thinking about people who aren't interested in competitions, or in winning and beating other people. The types who are more interested in creating things, and in open-ended discussions that do not have clear-cut results or resolutions. It would not make much sense to refer to that as 'feminine', and indeed the idea that it is somehow 'unmanly' may be one of the things that puts off more (straight) men from creative arts. This is one of the many problems we can avoid if we decouple character traits from gender (aside from the sheer fact that it doesn't make sense). Describing certain men as 'feminine' and certain women as 'masculine' suggests that there is something wrong with them. This is something that detractors use against Michelle Obama, and some take it even further and declare that she is not only 'masculine' (because she is strong, both mentally and phys-ically, exudes confidence, speaks up, and had her own career independently of her husband?) but that she is in fact a man (because surely only a man could have those characteris-tics?). This is a particularly egregious conflation of gender with character.

The issue of character and gender is often brought home to me when I'm teaching. While it can be useful for me to be aware of certain statistical trends, it is crucial for me to respond to each student in front of me as they are, not as the previous statistics might suggest. However, the previous statistics do suggest things I should be on the lookout for: girls who are convinced they are doing badly, when actu-ally they're doing quite well; boys who think they're doing brilliantly, when their proofs are actually full of holes. Girls who are too scared to speak up in class; boys who talk a lot and unremittingly dominate the conversation in the classroom.

I see similar trends among adults when I give public lectures. In one of my public presentations I typically ask if someone will do a juggling demonstration. Usually a man's hand will shoot up and he will stride happily to the front. Sometimes there is a long silence and things get a little uncomfortable. I plead with the audience to help me out because otherwise I will have to do the demo myself and it might not go well. Eventually then a woman might timidly put her hand up. She typically comes to the front and apologises, saying something like, 'Sorry, I'm not very good, I'm really out of practice, I haven't done this for ages.' And some of the timid women turn out to be much better at juggling than some of the confident men.

In question time there are typically many (usually white) men who instantly put their hand up to ask 'questions' which are often just criticisms thinly veiled as questions. Meanwhile many women quietly ask me genuine questions privately afterwards, saying they were afraid to ask in public. This trend is so marked that I have devised ways of taking questions after talks that do not involve anyone raising their hand or asking their question in public, about which more later.

However, all these trends are stereotypes and even if they are backed up by 'statistics' and 'averages', the sweeping statements erase individual experiences and ignore the fact that not all men overestimate their abilities and not all women are timid and scared. As a further nuance, these characteristics might be caused by gender – because of the way society treats men and women differently, they may well develop different levels of confidence or timidity in public – but this does not mean these are inherently male or female characteristics. So we would benefit from having ungendered terminology with which to talk about them, instead of saying that confidence is 'masculine'.

Ungendered terminology could then also help us think about how to encourage different characteristics that we wish to value and nurture in both men and women. In *Crossing* Deirdre McCloskey writes of her new experiences when she transitions to being a woman, and among other things finds that women will help her. Whereas, as she writes, 'female-to-male gender crossers must face the unhappy fact that American men don't help each other. The theory of American maleness is that your special woman takes care of you when you're sick, but aside from that you are supposed to do everything alone. Help among men is shameful, because it shows incompetence. Among women help is the point, because it shows love, "love" in its full sense: care, sympathy, providing for need.'

There is no reason why this needs to be a male-female split in the end, other than a deeply ingrained cultural one. I believe that if we adopt ungendered language to talk about these things then we can escape from those cultural restrictions, and everyone can be freed to benefit from the love and help of others.

Gradually, through all these experiences and thought processes, I came to realise that we need to think about new concepts, or think about old concepts in a new way. And I realised that new terminology would be a crucial key to thinking about these new concepts effectively.

We need new ungendered language in order to separate character traits from gender and have less divisive conversations in which people don't have to get defensive about 'not all men' or 'not all women' being a certain way. Because indeed men are not all the same, and nor are women. Not all men are aggressive, competitive, risk-taking and unempathetic. And even those who do behave in those ways might only do so because of social pressure and the idea, perpetuated by social norms, that this is how to be successful in society. When certain behaviour is rewarded by society many people will strive to behave in those ways even if at some deeper level it makes them unhappy.

Men and women currently suffer in opposite ways from these pressures: both may be unhappy with traditionally masculine behaviour, but if women decide not to emulate it they are at least accepted as feminine, whereas men may be criticised for a lack of 'masculinity'. For example, if a woman earns less than a male partner, or is younger, or less educated, it does not raise eyebrows nearly as much as if a man earns less than a female partner (or is younger or less educated). By contrast, if a woman takes on those 'masculine' characteristics, she may well be more successful but she may also be criticised because those same characteristics are seen as negative in a woman; where a man is ambitious and confident, a woman is pushy and aggressive. In a fairer society the pressures on men and women would be the same: whatever it is we value and encourage, we should value and encourage it from all people regardless of gender. This is not about being 'gender-blind' overall: we still need to see gender in order to fight directly gender-based injustice. It is about learning how to stop perpetuating subtler forms of systemic bias that favour men by favouring – for no good reason – characteristics traditionally associated with men.

My experience of gender imbalance and gender norms has changed dramatically now that I teach maths at the School of the Art Institute of Chicago. This is an 'art school' in the parlance of American education, that is to say, a university in which all degrees are in aspects of art: fine art, sculpture, architecture, design, fashion, photography, animation, and so on. I previously taught traditional maths students at traditionally academic universities, and my classes were often around 90% male. By contrast, only about 40% of the art students identify as male and, as you might expect, the male students do not really conform to any of the traditional 'masculine' character types. Some of them, especially the more straight-passing white ones, do carry with them some of the benefits of having grown up with society on their side, such as easier confidence when speaking up in group situations, but not all of them have benefited in this way and many of them are sensitive enough to the needs and voices of others that they tone this down in order to make sure others have a chance to participate.

It would be divisive, insulting, and I believe obstructive to say that these men are less 'masculine' than their aggressively competitive, probably more financially successful peers in the non-art world.

Mathematically speaking, if we have two things that are not equal we could make them equal by making the lesser one greater or by making the greater one less, or by a combination of both. However, there is a completely different way we could do it, which is by evaluating the two things on a new dimension entirely. We wouldn't necessarily make them equal in the end, but we would at least have made sure neither of them was affected by the original inequality.

To take an example, if two people were interviewing for a job as a typist we could equalise their heights by getting the shorter one to stand on a box. Or we could observe that height is irrelevant to working as a typist, so we should evaluate how well they type instead.

This sounds silly, but getting shorter people to stand on a box is essentially what we are doing all the time when we wonder about how and why men and women are different and what we should do about it. Instead, we could just understand people's relevant characteristics, which much of the time have nothing to do with their gender, instead of trying to use gender as a proxy for character. This doesn't mean that we are making men and women equal in a strict sense. Men and women are in fact not the same. What it means is that we are making sure that being a man or a woman doesn't have an effect unless it really has to (in a gynaecologist's office, for example). This is the idea of finding an entirely new dimension on which to think about things, one that is separate from that of gender.

A new dimension is a way of escaping the old dimensions. If you're on a train it is constrained by having to run on a track. That is one-dimensional (except where it branches) as the train can only go forwards and backwards, which is just the positive and negative of the same direction. If someone wants to block the train all they have to do is block in front and behind. But you could escape by invoking a second dimension: you could open a door and jump out.

At this point you would be moving in a left–right direction, which the train can't do.

Dimensions are directions, but they can be directions of ideas rather than of space. If we are thinking about choosing a restaurant for dinner, we might evaluate our choices according to price, cuisine, ambience, distance from home, and so on. Each of these is an abstract dimension, a dimension of ideas.

When considering dimensions of ideas, a new dimension still helps us to escape a trap; it's just that we're now escaping a trap of ideas. Perhaps this is the image that is being evoked if we use the now awfully clichéd imagery of 'thinking outside the box': we are using a new abstract dimension to escape limited low-dimensional thinking.

Sometimes the higher dimensions were there already. Life is complicated and we have to simplify it in order to understand it, and one way to do so is to depict it in lower dimensions like taking a picture, which is a two-dimensional depiction of our three-dimensional world. If we try and reconstruct a scene from a photo we're not exactly 'creating' a third dimension, but rather trying to rediscover the dimension that was eliminated by the photo. If we're dealing with dimensions of ideas, using a higher dimension consists of stopping the urge to squash out a direction of thinking. That sounds like it will make things more complicated, but sometimes it's only a short-term complication and can simplify things in the long run.

In Chapter 1 we talked about how maths uses a new dimension to escape the constraint of being unable to take square roots of negative numbers. Mathematicians came up with a new dimension of number which they called 'imaginary', and we represent it as a second dimension, usually vertical where the ordinary number line is horizontal across the page. The basic unit of this new direction is called i (for imaginary), where i is declared to be a square root of -1. Then the two-dimensional space that we can access by combining this direction with the old 'real number' direction is called the complex plane, and the points represent numbers called complex numbers which are a blend of real numbers and imaginary numbers. Here is a picture of the complex plane and some complex number positions on it:

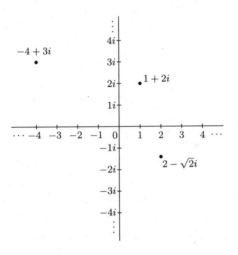

Although in some sense imaginary numbers were just 'imagined' by mathematicians, it turns out that this extra dimension helps us to see certain things more clearly. For example, in the real numbers the number 1 has two square roots, 1 and -1, because $1^2 = 1$ but also $(-1)^2 = 1$. However,

1 only has one cube root (which is 1), which is a bit unsatisfactory to mathematicians – why aren't there three, since we're now dealing with a power of three? In the complex numbers we can show[1] that there actually are three cube roots of 1, and moreover if we plot them on the complex plane we see that they are equally spaced around a circle of radius 1:

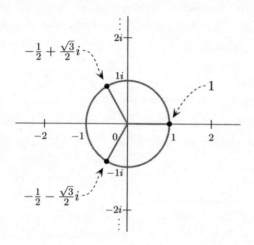

This geometric picture helps us to generalise to *n*-fold roots of 1, the complex numbers *x* such that $x^n = 1$ – these are also equally spaced around a circle of radius 1. Here is the picture for the 8th roots:

---

1 We can do some multiplying and show that

$$\left(-\frac{1}{2} + \frac{\sqrt{3}}{2}i\right)^3 = 1 \quad \text{and} \quad \left(-\frac{1}{2} - \frac{\sqrt{3}}{2}i\right)^3 = 1.$$

This is a calculation I would call tedious but not difficult.

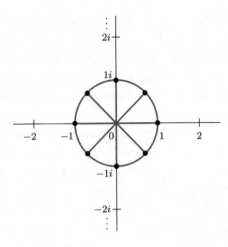

The picture and pattern are much more vivid in two dimensions; on the one-dimensional real line we can't really see anything.[2]

This is one example where using more dimensions is more complicated but also gives clarity and nuance, and is typical of what higher dimensions can do for us. The complication means that we will always be tempted to revert to lower dimensions for convenience, just as it is tempting to take a photo rather than try and remember all the details about a scene in your head. This is fine as long as we bear in mind that there are many different possible angles, and some angles will miss crucial details that will be captured by others. Being able to see things from different angles can

---

2 The whole picture is even clearer when using a polar coordinate expression for complex numbers (expressing points in the plane by a radius and angle), rather than the one we have here which essentially uses cartesian coordinates (expressing points in the plane by an X-coordinate and Y-coordinate).

be a decent substitute for actually working with the higher dimensions.

A great deal of my own research in higher-dimensional category theory involves moving between higher-dimensional structures and their lower-dimensional 'footprints'. The questions go in both directions: what sort of low-dimensional footprints will a higher-dimensional structure leave, and what can we deduce about higher-dimensional structures from their footprints? Often the footprints lead us astray and give us mistaken impressions about the higher dimensions, like the two-dimensional shadow in Chapter 1 which looked like a foot but turned out to be my hand.

I love the novel *Miss Smilla's Feeling for Snow* by Peter Høeg, and one detail that has always stayed with me is that Miss Smilla understands footprints in snow so well that she is able to reconstruct a child's last moments from the footprints he has left behind. She grew up playing a game in which children jumped around in the snow while someone closed their eyes, and that person was then challenged to work out what sort of jumps they did to produce those footprints. Sometimes this is what my work feels like.

Thinking about gender sometimes feels like this to me too: that we have taken a complex higher-dimensional structure of people's characters and encapsulated it in this one dimension in a flawed and reductive way, and our attempts to reconstruct the higher dimensions are not going very well. We need to do better.

It may seem that I'm simultaneously talking about finding a 'new' dimension and rediscovering one that we have squashed out of our consciousness. In maths the question of whether we invent things or discover something that was already there is a bit moot. Personally when I'm doing new research I feel like I'm *discovering* concepts that were already there, but *inventing* ways to talk about them. The latter aspect is important. Sometimes people say maths is 'just a language', as if language is a small thing. It's not.

Language is important. It can help us clarify our thoughts about something, even if we were already thinking about that thing before we had a word for it. Having a word for a concept is a way of making it more convenient to carry around in our brains. If you're not exhausted from carry-ing something around you can carry more things, take them further, and do more with them.

This is how we make new concepts in mathematics. Arguably the concept itself isn't new – it is there whether or not we give it a word. But giving it a word can open up whole new trains of thought, and enable us to build on the idea and carry it further.

One basic example of this is multiplication. Multiplica-tion of numbers can be thought of as just repeated addition, so 4×2 is

$$2 + 2 + 2 + 2.$$

In a way, multiplication is not a new concept and we don't really need a word for it, because it could be expressed as addition. But various new things become possible if we think of it as a new concept. First of all, we can do much more complicated versions, like 55×65, which would take a long time to do as repeated addition. But also we can now build a new concept from multiplication by repeating it in turn: $4 \times 4 \times 4$ gives us exponentiation, written 4^3. If we had to express this using only addition we've have to do

$$(4 + 4 + 4 + 4) + (4 + 4 + 4 + 4) + (4 + 4 + 4 + 4) + (4 + 4 + 4 + 4)$$

which is rather arduous. Furthermore we can think about versions of multiplication that aren't so much like repeated addition at all.

There's a concept of multiplying shapes together, multiplying patterns togther, multiplying symmetries together. This all began by thinking of repeated addition as a concept in its own right.

In this chapter I'm going to suggest new vocabulary to encapsulate some concepts that we may well have been thinking of all along. My aim is to help free us from all the problems we've discussed which arise from our conflation of character with gender.

New language

The word 'feminism' is gendered and thus divisive in many situations. I once bought *We Should All Be Feminists* by Chimamanda Ngozi Adichie and *Why I Am Not a Feminist* by Jessa Crispin at the same time in an airport bookshop. I thought this was quite a funny combination and made a comment about it to the checkout guy, but he gave no flicker of recognition whatsoever. I suppose I was just part of his job; I'm certain that if I had done this at my local family-run bookshop, Sandmeyer's, we'd have got into an interesting discussion about it.

As it turns out, those two books do not exactly contradict

each other, they just use different definitions of feminism. I have discussed the fact that some of the issues facing women are directly because they are women, and some are only indirectly so, and that there are many issues people face that are nothing to do with whether or not they are women. The divisive arguments over these competing issues – and the fact that they are in competition at all – act as a *de facto* tool serving the people who are already in power, and helping them to keep power.

Women who feel kept out of power are developing language to help them try and change that balance. That language may well serve to motivate and unite some women, and get them focused on what they want to achieve, but at the same time it too often alienates other people: men who might otherwise be allies, or more disadvantaged women who feel overlooked by the women who are themselves in positions of power among women.

One such sharply divisive word is the verb to 'mansplain' as coined by Rebecca Solnit. Women around the world breathed a collective sigh of relief and recognition when we were finally given a word for this thing that we feel happens to us often, where previously we couldn't quite put our finger on it or be sure we weren't just imagining it. I have written about straw person arguments against this concept before.[3] A few men have declared that they are unfollowing me on Twitter just because I use the word (I don't miss

---

3 Straw person argument: when you invent a different point that nobody is making, but that is easier to knock down, and then knock it down. But you haven't defeated anyone's argument, because nobody was making that argument. Previously known as 'straw man argument'. See *The Art of Logic*.

them). Some people declare that 'women mansplain too', showing that they have misunderstood what the concept is. To reiterate: mansplanation is not just when a man explains something to a woman in a patronising way. It is when he does it despite there being clear signs that the woman already knows it and possibly knows it better than he does, which fits into a pattern that many women experience across their whole lives, of society generally assuming that men are more knowledgeable than women. Thus women widely experience men discounting their knowledge and expertise, talking over them, claiming their ideas as their own, or even repeating what they've said immediately after they said it, as if it were a new idea. There is not a general widespread assumption by society that men are not experts in their field (though it does happen in some specific areas such as childcare), and so women by definition cannot engage in mansplanation. It's a bit like picking on someone smaller than you. If you're smaller than the other person, you cannot be described as picking on someone smaller than you. However, I acknowledge that this idea really winds some men up. Whether that matters or not is a separate question; either way it's an example of how divisive language can be.

Another word that women and feminists are increasingly throwing out is 'patriarchy', to acknowledge and draw attention to the fact that in our society as it currently stands, power is stacked in favour of men and against women. This word also riles up some men, often when they individually don't feel that they have much power in society. But the point is that far more men than women do hold power, and the *structures* of society favour men so that this skewing of the scales is deeply embedded. Some men in particular may feel that they have not benefited from the so-called 'patriarchy', for example gay men and poor men. One black male professor has told me that he thinks all women, including black women, are better off than black men in terms of the support they receive in life and the outcomes they experience.

For these reasons I try to qualify references to 'patriarchy' by calling it the 'straight white patriarchy' or the 'straight white rich patriarchy' or the 'straight white rich cis-patriarchy', but this leads me to wonder how many more adjectives we will have to attach to this situation. I do believe that it is important for us to take all these identities into account when considering the fairness of the world, notwithstanding some people, usually straight white men, rolling their eyes at 'identity politics'. But importantly I think we also need a way to think about relevant characteristics of people *separately* from the whole identity discussion. We need to stop blurring the issues. When someone's identity is relevant because of prejudiced attitudes of people around them or inequality embedded in society, then that is a genuine problem, but when we're thinking about character or abilities it does not need to have anything to do with gender or other identities. Whether or not it correlates with gender is not relevant, because correlation is not causation and does not completely determine what will be true about someone.

I have already talked about several of my experiences of women who do not fit the traditional 'feminine' type, including some of the professors at the women's college I was at, and the Five Billion Dollar Woman at the party. I have also seen character types that might be associated with femininity but include many men, such as all the artists not interested sport as I described at the beginning of the chapter.

To address our need to separate gender from character, I took a mathematical approach, not shying away from reorganising existing thoughts into new packages and giving them names. In maths when we name new concepts we sometimes use old words in new ways, if we are particularly seeking to make connections with previously known concepts. But sometimes we are trying to clear our minds and get rid of prior associations, and that's when we invent new words. That's also why we need new words here, to break our prior associations with gender and clear our minds.

I brainstormed for a year or two on and off with my dear, empathetic, emotionally sensitive, supportive friend Gregory, and came up with the following words: 'ingressive' and 'congressive'. The etymological idea is that 'ingressive' is about 'going into things', and 'congressive' is about 'bringing things together'.

These are descriptive words rather than prescriptive ones. They just describe behaviour, or attitudes, or situations. This can liberate people of all genders, and, in particular, men can be liberated from gender pressures as well. There might be a tendency to immediately constrain ourselves according to these new criteria, find ways to assess who is born ingressive and who is born congressive, but that is absolutely not the idea, as we'll see. Perhaps people feel drawn to putting themselves in categories just as much as they resist being put in them.

It is possible that these behaviours correlate with men or women in some ways, but that is a sort of afterthought, not a driving principle, and is in any case not an innate quality but one that is learnt through society's conditioning and so can be changed. It's possible that treating people fairly according to these characteristics will lead to changed outcomes in terms of gender equality as well, but the main point is that it will lead to changed outcomes in terms of *human* equality. We will still need to address sexism, racism, homophobia, transphobia, income inequality, and all forms of prejudice, bigotry and exclusion, but we will have greater clarity if we have better language to keep these issues separate from character traits. I will argue that these concepts and this terminology open dialogue and build bridges across identities rather than making our arguments more and more divisive and splintering us into more and more distinct and competing tribes.

With this ungendered language I can now rethink many of my past experiences and case studies with greater clarity. I can say that the Five Billion Dollar Woman was very ingressive, and this makes more sense than trying to say that she was very 'masculine' in her behaviour. Emmy Noether's work, and the work of the Bletchley women, was congressive. Dame Stephanie Shirley ran her business in a way that not only addressed discrimination against women directly, but also created a congressive environment in which congressive people could thrive. The books about women who are 'Headstrong' or 'Badass' are trying to portray women as great because they too can be ingressive. Sheryl Sandberg seems to be asking women to be more ingressive in order to be successful in an ingressive world. This certainly is one way to be successful, but we are going to examine other ways that we can approach more clearly now that we have this new terminology to help us.

Before we do that it is important to note that I am not claiming that this is a clean dichotomy. People aren't just ingressive or congressive. They can be somewhat one and somewhat the other under different circumstances. It's more of a two-dimensional plane, and we can be anywhere on it, and indeed move around on it across time and in different situations.

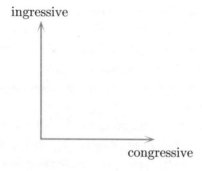

Importantly, these are not fixed characteristics but just types of behaviour, and like all behaviour they can be learnt. At a certain point I realised that I had spent most of my life learning to be more ingressive to be successful in an ingressive world, perhaps as Sheryl Sandberg suggests. But I didn't like myself that way. In my new portfolio career I have been unlearning that ingressive behaviour and finding that I can be successful while remaining true to my congressive self. Having the language to express it that way has helped me work out how to go about doing it.

Here are some fuller definitions or characterisations of these words:

ingressive: Focusing on oneself over society and community, imposing on people more than taking others into account, emphasising independence and individualism, more competitive and adversarial than collaborative, tendency towards selective or single-track thought processes.

congressive: Focusing on society and community over self, taking others into account more than imposing on them, emphasising interdependence and interconnectedness, more collaborative and cooperative than competitive, circumspect thought processes.

This not a clean dichotomy even though some aspects sound like exact opposites of each other. It's not a classification of people into two camps. It's a way to evaluate behaviour in a flexible and dynamic way to reflect the fact that people are flexible and dynamic, day by day and also over the course of their lives, not fixed and rigid.

Relationships with existing ideas

Although people are flexible and dynamic, the urge to separate their behaviour into (binary) gender categories is strong, and ideas based on those distinctions aren't completely useless. Those distinctions can be useful *to some extent*, but I believe the usefulness comes with many unhelpful side effects and that a new ungendered approach can preserve the usefulness while eliminating the side effects.

Indeed, most of what I read about gender issues seems to me to be really about ingressive and congressive issues, just without the terminology. And if it continues to relate the issues to gender, then it risks alienating those who do not fit the behaviour descriptions, and detracting from the point by causing arguments about whether it's really all men or all women, and whether it's patronising to simplify men and women in this way, and simplistic to make such a dichotomy.

Men Are from Mars, Women Are from Venus by John Gray is a famous and divisive example, but I actually found it rather useful. I didn't take it to mean that *all* men behave in a Mars-type way, but it helped me to recognise when someone (of any gender) was behaving in one way while I was behaving in the other, and thus helped me to communicate better and resolve situations that might otherwise have become ever more antagonistic. It helped me understand things like why I am resistant to advice when I tell someone about a problem in my life (because I am seeking validation, not a solution – in a Venus-like manner). I take the book to be really saying: 'Some people in some situations are from Mars, and others in other situations are from Venus, and it's often men who are one and women who are the other, but not necessarily, and you might be both at different times in different situations.' That is a somewhat less catchy title.

Somehow the imagery is much more vivid and arresting when it involves a very distinct dichotomy between two completely different things, but our need for something vivid and arresting then gets in the way of a nuanced understanding. Perhaps it's because we've been influenced to be too ingressive that ingressive dichotomies are necessary to attract our attention?

With the ingressive and congressive terminology we have a chance of avoiding those issues because we are focusing mainly on the characteristics, not the genders. It is a separate question to wonder whether men are more ingressive and women are more congressive, and if so, how much of that is learnt. But that is not the main question we will focus on. If we focus on the characteristics instead of the genders, we have more chance of seeing them as flexible and dynamic, both across people's lifetimes and day by day or minute by minute in different situations. Furthermore, as a somewhat abstract theory there is the possibility for shedding light on many different situations at once, across different parts of life, at different scales, and unifying many existing ideas about contrasting character types.

The Mars/Venus imagery largely focuses interactions in personal relationships, but there is also language aimed at professsional interactions. *Work Like a Woman* is the memoir by Mary Portas, who is also known as 'Mary Queen of Shops' and 'Queen of the High Street' after different TV series. Ostensibly she is writing about how she became successful and high-powered as a woman in a male-dominated business world. She also seems to me to be talking about ingressive versus congressive styles of business and leadership, without using those words. She talks about the 'alpha' and 'macho' culture of traditional business, which are good words but are still very strongly associated with men. She also talks about 'vertical' as opposed to 'circular' ambition and career paths. (Circular is not supposed to invoke the idea of going round and round in futile circles, but rather the idea of emanation and non-hierarchy.) I am slightly afraid that these words have too obvious a connection with gender; it's only a small mental shift from vertical and circular to phallic and yonic.

Jessa Crispin discusses related ideas in *Why I Am Not a Feminist*. She writes: 'The era of domination has to be replaced with an era of collaboration, not segmentation.' She calls on feminists to 'create a world of cooperation and fraternity', curiously using a male-gendered word at the last moment while essentially rousing us to build a more congressive world.

How new terminology helps us

Evidently I think that many insightful people have been writing about ingressive and congressive issues for a long time, just without the help of ungendered terminology. Here is how that terminology can work for people of all genders, as a new way of understanding the world, and perhaps even changing it for the better. These ideas come from my field of research, category theory, which has opened up huge new vistas of mathematical thought by giving subtly new points of view. One of them is to do with how we describe or characterise things.

There are often two ways to describe something: by some intrinsic characteristics or by the role it plays in a certain context. We could describe Barack Obama as male, a little over six feet tall, athletic, with short greying hair, or we could characterise him as the 44th president of the United States or the first non-white president. Each of those descriptions is true but accomplishes – and draws our attention to – different things.

In category theory, abstract mathematicians are very interested in describing things by the role they play in a context, rather than by their intrinsic characteristics. It is in a sense a more versatile way of thinking, as it means we are focusing on what is relevant to a situation. Sometimes something like hair colour might be relevant to playing a role – for example, if you are trying to be someone's body double. But in many contexts it is not relevant, such as if you're doing scientific research in a lab, or going up in a spaceship, or doing gymnastics – although all too often journalists still feel the need to describe women's hair when writing about them doing these things. Category theory says we should focus on what is relevant in any given context, which is determined by the roles that things play in that context, regardless of what their intrinsic characteristics are.

This is the approach I take when describing character types according to the new terms ingressive and congressive. It is about the roles that people and behaviours play in different aspects of society, not their gender, race, sexuality, wealth, or any other intrinsic measure.

I think this approach is fruitful and important in its own right whether or not it leads to greater gender equality, although I believe it will lead to that, as well as greater equality along other dimensions. I think that if we remove gender and other intrinsic identities from our arguments they can become less divisive and more productive. That means focusing on identity when it is relevant (which it often is, especially when we're thinking about prejudice) and not when it isn't. This amounts to separating out feminist arguments into the parts that really are about gender and the parts that are really about character, which might be more implicit or structural. We are separating out character and gender as independent variables, as we described in Chapter 3, instead of assuming that there is some formula relating the two. The existing approaches consist of continually trying to find 'formulae' for character based on gender, and continually finding that the formulae are inaccurate, divisive and dehumanising.

The best formula, really, is to admit that there isn't one. In reality the variables are probably neither fully dependent nor fully independent, so we do have a choice about how we treat them. If we treat them as dependent then we fall into all the traps of divisive arguments and prescriptive gender assocations. If we treat them as independent all the arguments become stronger, because we no longer need to pit genders against each other, but instead we ask what qualities of humans we are genuinely going to value in society. We can then work to nurture, value and reward those qualities in everyone of all genders. This is an important way in which individuals can influence structures, which can then influence individuals in turn.

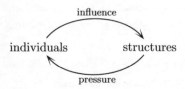

This is a cycle that is too often vicious, but we can work to make it virtuous instead. Sometimes we are obstructed in this by people denying that structural pressures exist. If an individual is personally prejudiced or oppressive towards another, that is one thing. But even if no individual is actively behaving in that way it is possible for society to have that overall effect on people, because of how society is structured. This difference between individual bias and structural bias is the cause of a great deal of antagonism and misunderstanding of identity politics.

Here is a diagram depicting some structural reasons why men continue to be more successful and powerful in the world, to help us understand what we are currently doing about it and what we could do instead. I am writing 'Y' for any general characteristics that are traditionally associated with men, such as self-confidence, ambition, competitiveness and risk-taking, but eventually we can consider Y to be general ingressivity:

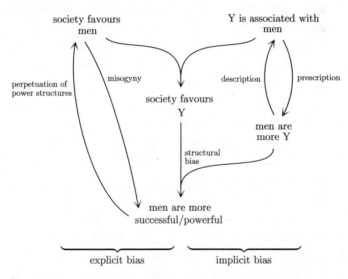

This diagram divides broadly into two parts. The left-hand side is explicit bias, that is, when society perpetuates the structural power of men through direct discrimination against women. This is the part that definitely needs to be addressed in a way that remains conscious of gender. However, if we're not careful this comes out as 'sexism towards men', although by some contemporary definitions of sexism it only counts as sexism if it's by the group that holds power (men) against the group that doesn't (women). In any case it is divisive because it's still based on gender and some people think this is now 'unfair' towards men in return.

I would call the right-hand side an aspect of implicit bias, in which society favours men not just because they're men *per se* but because of the way society favours certain characteristics that happen to be associated with men, perhaps unconsciously. I believe we are currently addressing the character bias in terms of gender, and this is causing other problems while also not fixing the old ones.

If we just address the fact that men are more likely to have the characteristic in question ('men are more Y'), this results in pseudo-feminism, or 'leaning in', the idea that women should learn those characteristics and 'become more like men' to be successful. Another approach tries to fix the fact that these characteristics are associated with men ('Y is associated with men') by doing what gets referred to as 'redefining masculinity' or persuading men to take on other characteristics that are more associated with women. Whether we are asking women to be more like men or men to be more like women, we are still basing our ideas on gender and remaining stuck in that one dimension.

But rather than addressing just the two sides of the diagram, there is a third way: the middle. We could consider why society favours those characteristics at all. This is the part that should not be based on gender. It is currently related to gender, but that is arguably for historical reasons rather than essential ones. I am proposing that we address that issue separately along our new dimension, which will detach it from both sides, that is, break the arrows that join it to the left and the right. When we break those arrows I believe we find that there is no particularly good reason for society to favour the Y characteristics at all, and we are then liberated to think about what characteristics we would actually like society to approve.

The new language helps us with this liberation. Instead of saying 'characteristics Y' we can say 'ingressive characteristics'. I argue that ingressive characteristics are associated with men, and men are currently more ingressive. Moreover, society favours ingressive behaviour, with the result that men are more successful and powerful. But I will also argue that *congressive* behaviour is actually better for society, so we should favour congressive behaviour instead. This will automatically change the gender balance of power in society without us having to take action along the gendered dimension.

How can we become more congressive, as a society and as individuals? Currently the most obvious way to be successful in our ingressive world involves becoming as ingressive as possible in order to be 'successful' according to those structures. But instead we can find congressive ways to be successful despite the ingressive structures of the world. Or we can change the entire structure to make the structure congressive. This second part of the book is about these possibilities for building a more congressive society. First, in the next chapter, I will describe ways in which I think society currently favours ingressive behaviour, for no particularly good reason.

5

Structures and society

Our current society broadly favours men, and I believe that one contributing factor is that society as it stands is largely ingressive. It does not mean that every individual in it is personally ingressive any more than it means that every individual is personally misogynistic. But I think it has led to general pressure to be as ingressive as possible in order to succeed, and that this has a knock-on effect on gender imbalances. In this chapter I'm going to explore how these ingressive structures are being perpetuated, ways in which they're bad for us, and what we could do instead.

As I am a mathematician my main experience of building environments and structures, and nurturing people to flourish, is in maths. I also think maths and science are key areas that are presented ingressively when they really have very congressive facets.

I have a particularly curious experience of this because of my unusual teaching experience: as I mentioned in the previous chapter, I taught traditional maths in traditional ingressive universities for some years, before switching to teaching maths to art students at the School of the Art Institute of Chicago. In the first part of my career I was teaching students who had done very well at school maths and who had enjoyed it enough to continue in university, and the classes were very male-dominated. I switched to teaching students who had typically not done well at school maths and often hated it so much that they fled from it as fast as possible as soon as they were allowed, and the gender balance flipped to being female-dominated.

Now that we know to go further back to more basic principles than gender, I can theorise that it is related to ingressive tendencies: that we teach and present maths ingressively, that this appeals more to ingressive people, and that this in turn means that it appeals more to men and puts more women off, although as usual if we talk about it via gender we have to say 'not all men' and 'not all women'. This also ties into my personal story, in which I made myself as ingressive as possible in order to be successful in an academic maths career, before becoming unhappy with that persona and trying to become more congressive, whereupon I discovered I didn't like the career any more.

In teaching my wonderful, intelligent, thoughtful, self-aware, often maths-phobic art students I have taken the opportunity to learn from them about their past maths experiences to find out what put them off maths, and crucially to develop ways to 'put them back on'. The short story is that they were put off maths by its ingressive nature and the ingressive way it is taught, and they are put back on it when I show them that it is actually very congressive, and when I present it in a congressive way. And I do believe that maths can be congressive and that education should be congressive.

However, like with so many aspects of society, it is very deeply ingrained that maths and education are ingressive, but I think this is yet another cycle of influences – it's ingressive because the rest of society is ingressive. Fortunately the School of the Art Institute is much more congressive than all the other places I've taught, in the trust it places in its teachers, the autonomy I am given, and the fact that I seem to be given liberating responsibility rather than oppressive accountability. So it has been an ongoing opportunity for me to move away from old ingressive teaching styles. I have been developing congressive ways to teach maths, and to develop a little utopia of a congressive society in my classroom.

My teaching is now unrecognisable from when I first started teaching at Cambridge University about twenty years ago. I was never exceedingly ingressive, but I was fitting in with an extremely ingressive environment and I taught in a very traditional way, as everyone else did: by standing at the front and writing a huge quantity of notes on a blackboard. The (mostly male) students frantically copied them down, tried to understand them later, and then took a gruelling exam at the end of the year.

Now I do almost exactly the opposite. Having the ungendered terminology has helped me to change this, because without it I might have been stuck over the question of whether I was trying to teach maths that would 'appeal more to girls', which sounds (and is) silly. There is a great deal of formal research and literature on different approaches to teaching and I am not claiming that my methods are new; indeed most of them are inspired by the best teachers I've learnt from in my life, either as a student or at conferences. The thing I want to stress is how much it has helped me to have the terminology with which to unify these various concepts in my head and also with my students. It means I can always ask myself, 'What is a more congressive way of doing this?' and when I think I didn't do something optimally I can often recognise that I was accidentally slightly too ingressive. I can recognise if a student is being a little ingressive and then realise that what I need to do is neutralise their behaviour and encourage them to be more congressive, with the result that everyone learns more, including them. Many educators work hard to make classrooms more congressive, whether or not they use the terminology. However, many classrooms are not very congressive, especially at university and especially in maths, because of basic assumptions about the ingressive nature of maths.

I'm going to describe how I have moved to congressive maths teaching, and discuss some ways in which I think it is misguided to think that maths is inherently ingressive. I will then look at how this principle of overvaluing ingressivity isn't just confined to maths, but is pervasive in education in general, and even beyond that to society at large. Education could be seen as the root of all these ingressive problems, but as always it's a cycle: education is ingressive because society is ingressive, and society is ingressive because we are all educated to be ingressive.

In *No Contest*, Alfie Kohn characterises competition as coming from situations where resources are scarce. But education involves a resource that can never be scarce: one person having knowledge and wisdom does not prevent someone else from having it. It might be scarce in the sense that not many people have it, especially when it comes to very specialised knowledge, but the whole point of education should be to share knowledge and wisdom with the next generation and thus ensure that it keeps growing. So the fact that we make education competitive is at worst contradictory and at best a choice that we should acknowledge and question. Ingressive behaviour is about more than competition, but this is a good place to start thinking about the assumptions embedded in our system.

Assumptions that go unquestioned stop us thinking about different points of view. A typical quip about mathematics is that 'one plus one just *does* equal two'. However, this is not exactly true. There are situations where one plus one is zero, such as if you say, 'I'm not not hungry', in which case you're really saying you are hungry: one 'not' plus one 'not' is zero nots. Or if you rotate something by 180 degrees and then rotate it by 180 degrees again, it's like doing it zero times.

There are also situations in which one plus one can be one, such as if you add a pile of sand to another pile of sand you still have one pile of sand, albeit a bigger one. If you mix one colour with one colour you get one colour.

If we focus too much attention on the supposed 'absolute truth' of basic arithmetic we risk getting into a mindset where we assume not only that adding things together makes more of them, but also the flipside: that using them or sharing them will make less. This is true of things like money, food and time, but there are other types of 'resource' that are not depleted by use. Knowledge is one such non-expendable resource, as are wisdom, curiosity, flexibility, love, joy. Not only are these not depleted by use or made smaller by sharing, but in fact they are probably depleted by *non-use* and increased by sharing. If you learn something but then don't use it for a long time then you are quite likely to forget it, whereas every time you use it you strengthen it, and one of the best ways to strengthen knowledge and deepen understanding is to share by teaching. Giving love, in a healthy situation, strengthens it and generates more love.

Education is a good place to investigate issues of competition and more general ingressivity because it can be examined at many levels – at the broad level of the entire education system, but also at more zoomed-in levels of individual schools and classrooms, classes, activities or methods. Every classroom is a small microcosm of society and teachers can influence those small microcosms very quickly and effectively, even if they can't affect the whole school or education system. It is also something that is a fairly universal experience, as we have all had some experience of education. Using the new ungendered terminology can help us understand that experience in more productive ways.

The first way in which the terminology helps me is that I start every course by discussing the theory of ingressive and congressive behaviour with my students and making it clear that I value congressive behaviour. In the classroom this means collaboration, contribution to the group, depth rather than speed, and curiosity rather than knowledge. Typically my art students are delighted by this and rapidly embrace it.

I also try to make the classroom as non-hierarchical as possible. The class is not about my power over students, it is about the process of mathematics. Importantly, I do not set myself up as imparting knowledge by authority, but rather, I am showing students what the processes of mathematics are as it is those processes that determine truth and value in maths, not my authority.

Mathematician and educator Prof. David Kung has a

thought-provoking TEDx talk[1] in which he points out that if we teach students to associate knowledge with authority rather than processes, then they will become adults who continue to do that and who will thus believe whoever they see as an authority figure, rather than the important processes of logic, evidence and reason. Essentially that is an ingressive approach to knowledge rather than a congressive one, and keeping that difference in mind can guide all our choices in how we set up structures in the classroom and beyond.

To help move away from a hierarchical structure I set up our desks in a U shape rather than the traditional rows with the 'lecturer' at the front. I also spend as little time as possible standing at the front doing traditional lecturing but instead do as much exploration and discussion as possible. I nurture, encourage and reward congressive behaviour such as curiosity, open-mindedness and collaboration, not ingressive behaviour such as showing off, posturing or belittling others. This also allows congressive people to reach their full potential without having to mimic ingressive behaviour.

Unfortunately many traditional clasrooms (especially at university) are ingressive, with all students – and the teacher – trying to show how brilliant they are, often by making others feel stupid. It is often boys doing it to girls and men doing it to women, but not always; it is ingressive people doing it to congressive people. A congressive classroom is instead one in which nobody is trying to win but everyone is trying to make sure the group is making progress and learning things.

---

1 David Kung, 'Math for Informed Citizens', TEDxGreatMills, 1 August 2019.

I give students projects to explore and investigate, where there is no right and wrong answer but a 'low floor and high ceiling', that is, a low barrier to entry and no real limit to what you can learn from it. I get them to collaborate rather than compete, with the idea that the whole class, between them, will contribute to the collective discovery process, nudged along by the teacher.

This expands to the possibilities for congressive outreach as well. Making maths into a competition or a game is thought to make it more 'fun', but it only makes it fun if you enjoy competitions and games. This risks putting off congressive people (including me). However, I love crafts and almost always prefer a maths craft to a maths game. I like maths fairs, where there are various tables of mathematical activities and children can spend as long or as little time as they like at each, with no particular aim. I am lucky that I was not put in for any maths competitions when I was a teenager as I would have hated it and possibly concluded that I was not cut out to be a mathematician. I worry that other congressive people have the same experience and that this leads to us shutting out congressive people from mathematics and, more broadly, science. Using this language, we can think more clearly about designing more congressive activities to appeal to more congressive people, without having to resort to saying silly things like 'activities to appeal to girls'. It may also help to explain why girls are so underrepresented in Olympiad teams, as we discussed in Chapter 3. If girls are being put off from entering the Olympiad just because they're girls, that is one thing to address, but if congressive people are being put off then I'd rather address that by building and valuing more congressive maths activities.

These sorts of activities, crucially, encourage depth rather than speed. Maths contains deep ideas and uses logical processes, both of which operate slowly. Much of the point is the process, and if we focus ingressively on getting to the outcome quickly then we will miss that point. There is a myth that maths is about getting the right answer, but that's a very ingressive approach to it; a more congressive approach, and one that is much more prevalent in higher-level maths, is that it's about how you construct rigorous arguments to show that something is an answer. A congressive approach says it's more important to learn that discipline than to learn the answer; it's also a more transferable skill. Maths detractors complain that they never have to use school maths in their daily lives, which is probably true if the whole focus was on specific answers about things like triangles and quadratic equations. Unfortunately, maths as presented at school is largely the ingressive aspects: getting the right answer, following rules that are imposed on you, a proliferation of facts that you're supposed to know, and solving problems. If instead the maths were congressive – about building arguments, seeing relationships between things, and understanding contexts in which different things are true – then the maths would be much more relevant to daily life and much more about genuine mathematical ability.

Some more ingressive people get on just fine with the traditional maths environments. I suspect that congressive people are put off school maths as it is presented so ingressively – in both the sense of presenting only the ingressive aspects of the subject, and presenting them ingressively. Furthermore, if boys are also pushed to be more ingressive than girls, then this would contribute to there being more men than women doing maths at university; conversely, art is more congressive in many obvious ways, and this might contribute to there being more women than men at art school. This is of course aside from the biases that are directly about gender, including discrimination, unequal expectations, stereotype threat, and the lack of role models.

I don't think we should make more congressive maths environments just to pander to congressive people, but in order to nurture congressive behaviour because of its value. I think it is valuable in life, and in particular in maths and science at all levels. A broad example of this comes from Finland.

Congressive education in Finland

The Finnish education system came into international prominence when a new large-scale assessment of secondary school students across the world was performed and the Finns came out top, surprising everyone including themselves, apparently. So global attention turned to Finland to see what it is they had been quietly doing.

Many surprises were in store. It turned out that the Finnish education system is essentially congressive. It focuses on collaboration, cooperation and student well-being, on equity rather than excellence, and values teachers and their autonomy. There is very little homework, the school days are shorter with many breaks, and holidays are longer.

This is very different from the previously lauded East Asian model of drilling, academic pressure, and huge quantities of homework from primary school onwards. It is described in *Finnish Lessons: What Can the World Learn from Educational Change in Finland?* by Pasi Sahlberg, director of the Finnish Ministry of Education's Centre for International Mobility. Important features he discusses include the fact that there is no standardised testing; a focus on cooperation not competition; free school meals, health care and counselling; not starting school until the age of seven; a generally relaxed and unregimented school day; the fact that there are no private schools in Finland (charging school fees is illegal).

A crucial part of all this is the system for selecting and monitoring teachers. Broadly speaking, teaching is a good job, with Finnish teachers having 'equal status with doctors and lawyers',[2] but more to the point teachers have professional autonomy, a curriculum that is only laid out in broad brush strokes, and a lot of support for children who need extra help. Teaching is thus a profession that many people actively want to enter. Potential teachers are chosen carefully, trained extensively (at the state's expense), and then allowed to have responsibility for their teaching instead of being saddled with accountability. Sahlberg says: 'There's no word for accountability in Finnish. Accountability is something that is left when responsibility has been subtracted.'

Previously, received wisdom might have said that this is all very well, but the students wouldn't learn as much; this possibility is contradicted by Finland having coming top in the world in the first three iterations of the Programme for International Student Assessment (PISA), which focused on reading, mathematics and science in turn.

---

2 LynNell Hancock, 'Why are Finland's schools successful?', *Smithsonian Magazine*, September 2011.

PISA is a wide-ranging system for testing a sample of 15-year-olds. It doesn't just test their proficiency: it also tests their motivation, attitudes and anxiety levels. In terms of proficiency, Finland came top in the first round of literacy tests, of mathematics tests, and also of science tests. This 'top' was rated according to which country had the smallest percentage of students performing at the lowest of six levels, but even if you switch to looking at which had the largest percentage performing at the highest level, Finland was only marginally outranked in each case – in literacy and science by New Zealand, and in mathematics by Korea and Hong Kong-China. The US was very far down each of the tables. Unlike Hong Kong, Finland also has one of the smallest differences between the strongest and weakest students (while still having one the highest averages) and one of the lowest correlations between family background and eventual proficiency (unlike the US).

There are plenty of people who are so conditioned into the ingressive style of education that they refuse to admit that the Finnish system could work anywhere else. They insist it must be some sort of anomaly, only possible because Finland is so small and homogeneous, or because of some sort of Finnish genetics, or because the weather is so bad that nobody does anything except sit at home doing their homework.

But Finland is only as small and homogeneous as many states in the US, where education is run at the state level in any case. Moreover, Finland has had an influx of immigrants in recent years, meaning that some areas have become a lot less homogeneous than others, but the differences in homogeneity don't seem to have much impact on the schools' outcomes. And they don't really have homework.

Incidentally, Sahlberg points out that they're not that interested in their PISA success, as they're more interested in educating children to learn, not to take a test. They're more interested in equity than excellence, but it's funny how they focused on equity and achieved both greater equity *and* greater excellence, where other countries that focus on excellence achieve neither. Focusing on the good of society resulted in more highly proficient individuals as well.

There is a slightly sad sequel to the Finland story, which is that after becoming the focus of international attention Finland's PISA scores started dropping, leading some people to gloat that it was all a mirage or a fluke. (Still, at 12th place in maths in 2015 it remained far ahead of the US, which was in 40th place.) Sahlberg gives various highly plausible reasons for this, including a sort of feedback loop – a system that did not previously aim for success in tests did well in some tests and then became self-conscious about it. Sahlberg says that the country lost the drive to keep improving the education system, whereas other countries began deliberately aiming for higher PISA scores. Nevertheless Sahlberg maintains a congressive approach and says that the best thing to do is not adjust the system to aim for higher PISA scores, but to enhance equity and equality (which have also dropped in Finland's recent results). He says: 'The Finnish way of thinking is that the best way to address insufficient educational performance is not to raise standards or increase instruction time (or homework) but make school a more interesting and enjoyable place for all.'[3]

---

3 Joe Heim, 'Finland's schools were once the envy of the world. Now, they're slipping', *Washington Post*, 8 December 2016.

How this relates to gender

One way in which Finland still sets itself apart from almost all other countries is in gender equality; I suspect that many problems around gender bias in other countries can be traced back to an ingressive education system designed, unconsciously or otherwise, to promote inequality.

Most countries have boys doing better than girls in maths and science, and girls doing better in reading, but Finland has girls doing better than boys in maths and science as well, according to various measures. Digging into the specific global results yields interesting details, though.[4] Science proficiency is divided into different categories and gender differences are measured in each one. Boys do better than girls on average overall, but not in every category. Boys do better in content knowledge, but girls do better in 'procedural and epistemic knowledge'; boys do better at 'explaining phenomena scientifically', but girls do better at 'evaluating and designing scientific enquiry', and slightly better at 'interpreting data and evidence scientifically'. It seems that girls are doing better at the more congressive aspects of science.

In fact, there is also data on gender differences in students' epistemic beliefs. The students were asked how much they believe in things like 'The ideas in science sometimes change', 'Sometimes scientists change their minds about what is true in science', and, rather crucially, 'It is good to try experiments more than once to make sure of your findings'. Girls were found to believe in all these statements slightly more than boys. Again, I think these are congressive

---

4 See PISA 2015 Results, vol. I, OECD Library.

aspects of science that are perhaps often neglected by the general public and conventional presentation of science.

The last statement in particular seems related to the idea of self-confidence. If you are overconfident then you are less likely to check your answers and indeed your experiments. The PISA study even provides scope for investigating this as, along with assessing students' proficiency, it includes questions about students' attitudes. The results on 'self-efficacy' are fascinating.[5] This is essentially a measure of self-belief, as students were asked questions about how easily they think they are able to do certain things. It will surprise few people that boys turned out to have more confidence in their ability than girls, and it's tempting to conclude that self-confidence leads to higher achievement, but we must remember that correlation does not imply causation. In fact it is to be expected that those who are better at maths and science should believe that they are better and that those who are worse should believe that they are worse. The fascinating thing is that boys have more self-belief than girls even in Finland, where girls do actually do better than boys. Moreover, although self-belief did mildly correlate with proficiency within countries, it did not do so across countries: some of the highest levels of self-belief were reported in the US (which does not actually do very well in terms of proficiency), and the lowest were reported in Hong Kong-China, where there were much higher levels of proficiency overall.

This was studied even more specifically by three researchers,[6] who used some questions specifically to test

---

5 See p. 137 of the report.

6 J. Jerrim, P. Parker and N. Shure, 'Bullshitters. Who are they and

'bullshitting' in mathematics. In a marvellously satisfying ruse, they asked students to rate their own knowledge of sixteen maths concepts. The ruse was that three of the listed concepts were fictional: 'proper number', 'subjunctive scaling' and 'declarative fraction' (the questions were only put to students in English-speaking countries). Thus anyone who reckoned they were very knowledgeable about these things was deemed to be 'bullshitting'. Again it will surprise nobody that boys were found to bullshit more than girls, but the fascinating thing was that in North America the gender difference was much less – apparently American and Canadian girls have learnt to bullshit just as much as their male peers, but in Europe only the boys have learnt to bullshit as much as the Americans.

Of course, some aspects of these gender differences really are to do with the different ways in which society treats men and women: if men are told they are brilliant a lot, then it's no surprise that they believe they are brilliant. If society and in particular men are constantly putting women down, then it's no surprise that women will not believe in themselves. But if we use ungendered terminology we can separate that out from the question of what we actually want to encourage and nurture, which is particularly important when we're thinking about education, both at the level of broad educational systems and at the level of individual schools and classrooms. And then we can ask more careful questions about which characteristics are beneficial.

Self-confidence can benefit or harm students in different

what do we know about their lives?', *IZA Discussion Papers* 12282, 2019.

contexts because it can work in opposite ways. It can help students continue when they're struggling, which is important, but it can also enable students to overestimate their abilities, so that they don't work hard or seek help. It can lead people to jump to conclusions and not check their reasoning – in normal life this means that people do not fact-check, in scientific research it means people do not check their data or their methodology, and in mathematics it means people don't check their argument. This happens at all levels of maths education, from children who do their work quickly without checking and so have many wrong answers, to researchers who write down that things are 'clearly' true but then nobody else can figure out why that is the case; and sometimes it isn't.

So how much self-confidence is good and how much is counterproductive? If we look at successful mathematicians and scientists we find a variety of possibilities.

The role of self-confidence

The extraordinary and brilliant Indian mathematician Srinivasa Ramanujan came to wider attention outside mathematics following the film about his life based on the book *The Man Who Knew Infinity* by Robert Kanigel. Ramanujan failed his (ingressive) high school exams and failed to get into university several times but persisted in pursuing research by himself and eventually became recognised as one of the greatest ever mathematicians, albeit one with unconventional methods from the rigid point of view of European academia. His break came when Cambridge mathematician G. H. Hardy was able and willing to recognise the insight in

Ramanujan's work in its own right, independently of external (ingressive) measures like grades and degrees.

In order for Ramanujan to persist in his research despite multiple external setbacks, he had to have a certain personality, a huge belief in himself and the conviction that he was brilliant and that the world was just not recognising him yet. I have met other (much less brilliant) mathematicians who have this conviction despite multiple failures at exams, grad school applications, job applications, and they have all been men. In my experience of helping students, women are much more likely to conclude after one initial setback that they are not good enough. Of course, it doesn't help that they may have men all around them explicitly telling them they are not good enough (even when they are). And of course, it's not just women: it's congressive people.

The obvious conclusion to draw is that self-belief and self-confidence are valuable assets for a mathematician. But there is, as usual, something more nuanced going on. There are large numbers of people out there who really aren't brilliant at all, who fail according to all external measures because they aren't brilliant, but who still persist in believing that they're brilliant and that the world is just unfairly not recognising them yet. Prominent physicists and mathematicians receive a huge quantity of unsolicited mail from these types of people, who are convinced they are great geniuses who have solved everything, despite the fact that they have no job, no research record, and possibly no degree or training in the subject. It is self-belief and resilience to failure taken to an extreme: ingressivity taken much too far. I have only ever received one such letter from a woman, and the rest have all been from men. Physicist John Baez calls them 'crackpots' and has devised a 'crackpot index' with which to assess how far gone they are. It is tongue-in-cheek but also has a large element of truth in it.

At the other end of the scale there are congressive students who are so aware of their own weaknesses that if they don't come top they might give up, believing that they're not good enough. They are the kind of students who don't imagine they're good enough to go to university, or don't imagine they're good enough to do a PhD, or don't imagine they're good enough to have a career as a mathematician. If I hadn't got my first-choice PhD place I would have given up – I had fully decided that I would go and do something else instead because that would be a sign that I wasn't good enough. Note that this isn't just imposter syndrome – for women and minorities it might be because people have actively been doubting them all their lives, even when they have external markers of sucess to show.[7]

One possible conclusion is that we should get everyone to be more ingressive so that they don't give up. But then we'll end up with more people not giving up *who should give up*, and we also risk making congressive people feel uncomfortable or perhaps being put off by the very idea that they have to be more ingressive to succeed. Another possible argument is that we should get people to have a better balance between ingressivity and congressivity so that they don't give up so easily but also don't persist unrealistically.

But I am going to argue that there are important ways in which congressivity benefits *individuals*, that they would

---

7 Thanks to mathematician Dr Marissa Kawehi Loving for drawing my attention to this exact subtlety.

lose that benefit if we managed to get them to be less congressive, and that the benefits of ingressivity are things we could replace with structural measures so that people don't lose out if we encourage them to be less ingressive. I would rather work with or teach someone who underestimates and doubts their own abilities than someone who overestimates them.

Like with the 'dandelion' and 'orchid' children, I would rather build congressive environments and structures to help congressive people flourish, instead of trying to get them all to be more ingressive. In the case of congressive students, one way in which we can help is to find the ones who underestimate their abilities, and provide them with the validation they need in order to keep pursuing what they're trying to do. The only reason that their congressivity is a hindrance is because the people around them do not provide that kind of support. Besides which, some people might want (or be able) to become more ingressive.

This is a way in which we can address gender imbalances directly, by specifically finding women who underestimate their own abilities and who may have been put off by peers or professors belittling them; I am grateful for the support I have received in this respect. But it doesn't just have to be about gender. I think any congressive people have the potential to be better mathematicians, scientists, and general contributors to society if we nurture them well. In fact, many brilliant scientists talk about their own self-doubt.

Congressivity making scientists better

Dame Jocelyn Bell Burnell is an astrophysicist whose PhD work resulted in a Nobel Prize – for her supervisor (who happened to be male). She was not bitter about it, acknowledging that supervisors usually do get credit for the graduate work done under their supervision, and that Nobel Prizes should not go to graduate students. She has proceeded to have a stellar career and in 2018 was awarded the Special Breakthrough Prize in Fundamental Physics for 'fundamental contributions to the discovery of pulsars, and a lifetime of inspiring leadership in the scientific community'.

It seems that her 'fear of flunking made her meticulous'.[8] I read this as congressivity improving one's rigour as a scientist. Before I submit a paper for publication the fear of a referee finding something wrong with it makes me check it even harder. Before giving a talk the fear of being challenged by someone in the audience makes me prepare much better and anticipate all possible questions. Incidentally the Special Breakthrough Prize comes with $3 million, significantly more than the Nobel, and Bell Burnell donated the entirety of it to help female, under-represented ethnic minority and refugee students become physics researchers,[9] an extremely congressive act.

Large prizes are a rather ingressive way to celebrate people, especially when the prize has to be awarded to individuals rather than groups. In his memoir *The Gene Machine*, Nobel Prize winner Sir Venki Ramakrishnan points out how out of date the Nobel Prize is in its self-imposed limit of naming only three winners. That was not unreasonable at

---

8 Sarah Kaplan and Antonia Noori Farzan, 'She made the discovery, but a man got the Nobel. A half-century later, she's won a $3 million prize', *Washington Post*, 8 September 2018.
9 Pallab Ghosh, 'Bell Burnell: Physics star gives away £2.3m prize', BBC News, 6 September 2018.

a previous time in the history of scientific research, when collaboration was slow, unwieldy or impossible because of the difficulties of communicating across long distances. But the ease and speed of modern global communication has enabled collaboration with no apparent limitation on distance or scale, and the three-person limit seems increasingly absurd. The 2016 detection of gravitation waves involved over a thousand scientists across the world, but the Nobel Prize could name only three: Rainer Weiss, Kip Thorne and Barry C. Barish. A 2015 paper on the Higgs Boson has an astonishing 5,154 authors listed; the paper consists of nine pages of research together with 24 pages of authors' names and institutions.[10]

Mathematics can still be done alone, although it is increasingly done in collaboration. Fields Medallist Terry Tao talks about how he was the ultimate ingressive mathematician when young (although he doesn't use that word). He was competitive and a classic 'prodigy' who sailed through levels of education and excelled in competitions. But in his adult life he has become extremely collaborative and has written papers with an extraordinary hundred or so people, as of 2019. In one case he set up an open online collaboration that made rapid progress on a problem as 'people with different skills squeezed out what improvements they could'.[11] The activity was sparked by a result

---

10 G. Aad et al. (ATLAS Collaboration, CMS Collaboration), "Combined Measurement of the Higgs Boson Mass in pp Collisions at $\sqrt{s}=7$ and 8 TeV with the ATLAS and CMS Experiments," *Physical Review Letters* 114, 191803, 14 May 2015.

11 Erica Klarreich, 'Together and alone, Closing the prime gap', *Quanta Magazine*, 19 November 2013.

obtained by a very different style of mathematician, Yitang Zhang, who was almost unheard of, and who more closely fits the old stereotype of a solitary genius toiling away obsessively by themself. Zhang is quoted as saying, 'I choose my own way, but it's only my way',[12] whereas Tao is somewhat more direct about discouraging young mathematicians from that route. In another congressive facet, he is a prolific and generous blogger, and in one post wrote about the solitary, obsessional path as a 'particularly dangerous occupational hazard'[13] and in another advised being very sceptical about your work, most of all if you're dealing with a difficult problem.[14]

This is sheer speculation, but perhaps Maryam Mirzakhani is another example of a mathematician who was fast and competitive when young and then became 'slow' in order to be a deeply brilliant researcher as an adult. The skills needed for research are not the same as those required for Olympiad-like competitions or even exams. Research involves collaboration, reference to all the existing literature, and doing things nobody knows how to do yet, in a time frame, usually, of years. There are many examples of great researchers who were not good at exams. I think our focus on exams and competitions is missing the point, and is sometimes actually harmful.

One example where competition in mathematics was obstructive was the long-running rivalry between Newton

---

12 Ibid.
13 Terry Tao, 'Don't prematurely obsess on a single "big problem" or "big theory"', terrytao.wordpress.com.
14 Terry Tao, 'Be sceptical of your own work', terrytao.wordpress. com.

and Leibniz about calculus in the late seventeenth and early eighteenth century. They had different approaches and were fiercely competitive about whose was first. Arguably Newton's insistence on his supremacy, and his extremely revered status in Britain, held back British mathematics for a hundred years as British mathematicians were stuck using his notation for differentiation where Leibniz's was much better.

What if we pooled our knowledge and collaborated instead? In small fields of mathematical research that is a much more common way to deal with different people working on the same thing in different parts of the world. This has become especially widespread now that communication is so easy and fast.

In my field, category theory, I have never seen a mathematician race to beat someone at their research. If someone is already working on a problem we are more likely to leave it to them to avoid that type of competition, or collaborate with them if we think we can help. Physicists and mathematicians have been sharing their work online for free more or less since the technology to do so first existed. We post articles online long before they are peer-reviewed, and this helps us get wide feedback which in turn helps us improve our work. This is a congressive way of doing research. Keeping it secret until a big shock announcement is ingressive.

Many congressive mathematicians blog about their work almost in real time, sharing it widely while it is going on in order to get as much collaborative input from the rest of the world as possible. In fact, in one contemporary field related to category theory, homotopy type theory, an entire book has been collaboratively written using online technology, and it seems that the whole process, including dissemination, has been deeply congressive; one of the (over forty) authors, Andrej Bauer, wrote on his blog that 'the spirit of collaboration that pervaded our group...was truly amazing. We did not fragment. We talked, shared ideas, explained things to each other... The result was a substantial increase in productivity. There is a lesson to be learned here.'[15]

The general idea that sharing things makes them less is an ingressive mindset that leads to competition and individualism. I have been looking at maths and science, in both education and research, but this applies more broadly. Many other structures of society are set up as if they are all driven by scarcity, when, as with education, there is no real scarcity. I am going to argue that these structures are based on assumptions that ingressivity is better, and that we thus think that filtering people ingressively will select the 'best', when in fact congressivity might be much better.

I argue that ingressiveness is not actually better for society and so society would be better off not favouring it. If we try and fix gender issues without addressing this, then the fix will be superficial, like increasing diversity without improving inclusion.

Ingressive structures

Society favours ingressive behaviour structurally, in the form of systems that we use to run things, and filters that we use to select people for those things. This can happen because

15 Andrej Bauer. 'The HoTT book', math.andrej.com, 20 June 2013.

of individuals who consciously favour ingressive behaviour, but it can happen even when they don't. I have found myself accidentally defaulting to ingressive structures, and on one occasion it was actually when I had just given a talk about ingressive and congressive traits at a big conference.

There were several hundred people in the room, and at the end I announced that I had forty preview copies of my book to give away. There was a gratifying stampede of people trying to get to the front to get a free book, and then a gratifying moment of sheepishness as they all realised that they had been rather ingressive about getting a free book ahead of everyone else, while the more congressive people had stayed at the back. I started wishing I had come up with some sort of congressive method for handing out the books – perhaps an invitation to people to come and pick up a book to give to someone in the room who they felt really deserved it – but I had been rather congressive myself and not imagined that so many people would want the book.

Anyway it brought into focus how often in life we reward ingressive behaviour, even accidentally. And even if we do it deliberately we might not have done it *consciously*. Ingressivity is so ingrained in our upbringing, in our education, and in society that we might just think we're rewarding something that's 'good', such as with the 'strong' arguments being favoured in Oxbridge exams, without stopping to think about whether there's a congressive version of 'good' that we're overlooking and that might even be better. At other times we might just never have thought about it, such as in 'first come first served' situations. For example, when people raise their hand to ask questions, why does the person who raises their hand first deserve to ask the first question? As mathematicians we challenge the apparently basic principles and try to find more basic ones.

The fact that ingressive behaviour is better for oneself and that congressive behaviour is better for society is more or less built into the definition. But, as with most things, there is more nuance to this point. If you accept that society is made up of a collection of individuals, then the character of society broadly comes from two things: the characters of the individuals and also the structures they have for interaction. But individuals' behaviour is in turn affected by society.

Perhaps there is a cycle of interactions like this:

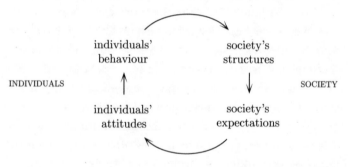

If we accept that society has an effect on individuals, then we see that what is good for society is also good for the individuals who are part of that society. In that sense, congressive behaviour is good for society *and thus also* good for individuals in that society, whereas ingressive behaviour can be good for individual people in society without being good for the whole. It is then a frustrating fact that although congressive behaviour is better for society, our society is set up to reward ingressive behaviour.

One ingressive filter that is very commonplace in society is competitions: explicit competitions with prizes, and competitive structures such as exams and elections. We use them even for congressive roles such as research, teaching, management and playing music. Arguably, running a country also ought to be a congressive role, but we still select for it using ingressive systems; I'll come back to the ingressivity of politics in the last chapter.

Through ubiquitous competitions we put pressure on people to be competitive if they want to succeed, or to give up on success if they don't want to be competitive. This starts in education, as we've discussed, but continues all across different aspects of society. It is no surprise, then, that people do tend to be competitive, but this does not show that such behaviour is innate or 'natural'. I'm going to examine our assumptions around competition in a way that we could apply to any aspect of ingressive behaviour. We question whether or not it's innate, what cycles are causing us to favour it, and what we stand to gain if we change our assumptions and our structures that perpetuate it.

In *No Contest*, Alfie Kohn conducts a comprehensive review of a vast quantity of research on competition. He debunks many myths, including the idea that competitive behaviour is innate, that women are naturally less competitive, and that competition is even good for anything at all.

As usual, one way to study innateness is to look at children's behaviour. My nephew, when he was about five, asked in that earnest way children do, 'Why did they have to have a World War? Why couldn't they just have a World Meeting and sort it out like that?' And apparently it's not just him. Alfie Kohn tells of various ways in which children are observed to prefer collaborative games to competitive ones. If children are exposed to competitive games they learn to become more and more competitive, but Kohn finds no studies showing a preference for competition once children have experienced cooperation. Preference is not the same as innateness, but if children prefer to be cooperative this does throw doubt on the idea that people are 'naturally' competitive.

Of course, children are exposed to competitive games all around them. In many cultures, particularly in the US, they are encouraged into competitive sport in a way that they are not generally encouraged (in such a ubiquitous way) into collaborative music, theatre, or other creative arts. One justification is that it's a way to pay your college fees (at least in the US), but then the question remains: why is it possible to pay your way through college by playing competitive sport, but not by music or theatre?

Note that there are scholarships for musicians to study music, but the difference with sport scholarships is that you can get a sport scholarship to pay for your degree in *something else*. Why can't you get a music scholarship to pay for your maths degree?

We could answer pragmatically that it's because sport generates far more money than music or theatre, so universities make much more money from it and thus they can pay the student participants more and, most importantly, want to do so in order to attract the best sportspeople away from other universities.

Let us keep questioning this system to try and get further back into first principles: why does sport bring in so much more money than music or theatre? Because more of the general public enjoy watching it, and either pay for it directly or contribute to huge home viewing figures which then generate advertising revenue. Why do people enjoy it? Well that comes back in a big circle to them being exposed and pushed towards it as children.

Once we see we're caught in a cycle we can ask ourselves whether it's a good cycle or not. As it's not based on anything, we could consider breaking it. In the last chapter we will come back to ways in which we can change structures to be more congressive; in this case it would mean building in structures to offset the pull of competitive sport as a way of paying for a university education. The most obvious congressive structure that addresses this is having more affordable or even free public education, which is probably why sport scholarships are not such a skewing influence in Europe, where university education is often state-funded. We've come back in a circle to education again.

Contrived scarcity

As we have described, Alfie Kohn characterises competition as something that happens when resources are scarce. From here we get the idea that competition is a 'natural' behaviour, because 'in nature' resources are scarce and so we have to compete for them. The thing is that, as with education, in many situations in modern life there isn't any real expendable resource at play, but a scarcity is fabricated in order to create artificial competition. This seems to be not only because we believe competition is good, but also because we believe people simply enjoy it.

And so things are made into a competition by contriving a scarcity. Children are put into music competitions in order to rank them even though playing music well is not a limited resource. They are put into maths competitions, spelling competitions, general knowledge competitions. It is said that this makes it 'fun'. But it only makes it fun for children who like competition, and plenty don't, including me.

A vivid example is the children's game Musical Chairs, in which you deliberately start with one fewer chairs than children so that children have to compete for the chairs. At each stage you remove the 'loser' and you also remove one further chair so that you contrive scarcity for the next round as well.

In adult life this manifests itself in all the singing, dancing, baking or 'survival' TV shows in which one person is kicked off the show each week, until there is only one winner. This is a form of contrived scarcity because 'winning' is not a genuine scarcity – the only reason there can be only one winner is that someone has decided it to be so. The scarcity is contrived rather than real for most prizes, unless the point of the competition is, say, to select an architect for a particular building project.

In 2019 the Booker Prize was notoriously awarded jointly to Margaret Atwood and Bernardine Evaristo. This caused some general outrage because that was 'against the rules' or because 'failing to pick one winner shows you have failed'. Failed at what? Well, you've failed at picking one winner, but what was the point of picking only one winner except that the rules said so? Where did the rules come from? Charlotte Higgins wrote in the *Guardian*: 'Everyone agrees that competition is the enemy of art. And yet, on the whole, there is also an agreement to conspire in the notion that it isn't.'[16] She gives some examples of recent prizewinners who

---

16 Charlotte Higgins, 'The Booker prize judges have exposed

shared their prize with their fellow nominees: artist Helen Marten (the Turner Prize), and writer Olivia Laing (the James Tait Black Prize). In fact in 2019 the four shortlisted artists for the Turner Prize requested to share it before it was even awarded, and the jury agreed. How does this relate to gender? Is it significant that two of the early winners to share their prize were women? Either way, I would say that this is a very congressive act, and Higgins ends her article by predicting that more artists and writers will take this congressive approach to prizes in the future, and suggesting that prizes themselves have the option of adapting to this congressive mood or losing their cultural dominance. She doesn't use the word 'congressive', but I see it in this light. If I said it was a feminine thing to do, then that would do an injustice any congressive men who might feel inclined to do similarly.

So much for the idea that competition is an inevitable product of society. What about the idea that competitive behaviour is genuinely valuable? If we see that successful people tend to be competitive and that uncompetitive people tend to be less successful, we might conclude that competitive behaviour is beneficial for success in practice. However, this is one of those situations where there are more root causes contributing to that outcome of success, and if we think harder about first principles we can uncover them. In this case, it's that society's structures are set up to reward competitive behaviour, so by definition competitive behaviour will yield greater success.

the doublethink behind our arts awards', *Guardian*, 15 October 2019.

Is it fair to reward competitive behaviour in this way? It is instructive to challenge the idea that competition is beneficial. There are many ways in which competition could be actively harmful, and there are also many ways in which different motivations are possible, so that whatever we believe is achieved by competition could actually be achieved in some more congressive way. Finding those congressive ways would have the double effect of avoiding the harm caused by competition and at the same time being inclusive towards congressive people; we are also going to see ways in which congressive approaches might be positively advantageous.

It might be tempting to point to many discoveries and achievements that were driven by fierce competition, including in science, the space race, climbing Mount Everest, reaching the South Pole, and so on. But it's important to balance things out in these discussions, and weigh up the ways in which competition and collaboration might both contribute to and hinder progress, rather than simply pointing out things that have been achieved through competition. Could those things have been achieved without competition? As I mentioned earlier, my field of research, category theory, does not really involve people competing with each other for research. Many scientific discoveries came about by collaborative work rather than teams competing to be first: the recent first ever image of a black hole, for example. This necessarily involved scientists and telescopes around the world: as the black hole disappeared over the horizon for one telescope, others started picking it up, and the images were then 'stitched' together by supercomputer. The network of telescopes is called the Event Horizon Telescope and was founded by astronomer Shep Doeleman. More telescopes have joined in since that initial picture was created.

So competition is not always necessary for scientific achievement, and moreover competition can have bad side effects. One meta-analysis[17] found that almost 2% of scientists admit to having 'fabricated, falsified or modified data at least once' and almost 34% admitted 'other questionable research practices'. We should wonder whether they are driven by the competition to publish, which is in turn driven by the ingressive way scientists and institutions are evaluated by how many papers they have published.

When universities compete in a fee-driven market this arguably results in more expensive facilities and grade inflation, not necessarily better education itself (unless you think facilities and grades equate with quality of education). The competition among students to win places might have led some students to work harder, but it has also led some parents to employ more and more nefarious means to get their children a 'prestigious' university place.

There is also a question of whether individuals are

---

17 D. Fanelli, 'How many scientists fabricate and falsify research? A systematic review and meta-analysis of survey data', *PLOS One*, vol. 4, no. 5 (2009), e5738.

spurred on by competition to do better, or not. Highly competitive situations can lead people to cheat, deliberately sabotage others, or give up. They might give up because they can never win (like me and sport), but they might also give up after the competition has ended because they did not develop any intrinsic motivation for the activity, only the extrinsic motivation for winning.

I am an example of someone who does not do better in competition, but rather in collaboration. I am sure that there are many doctors and nurses who are motivated by wanting to help sick people, not just by competing for prizes or income. I am sure there are firefighters who are motivated by saving lives, not trying to compete to save the most lives. What about teachers who are motivated by wanting to educate young people, and musicians who keep working to do better and better even when they're not in competition with anyone? There are men and women working in all these fields, but research shows that men tend to be more competitive than women. For example, one study in the *Quarterly Journal of Economics*[18] found that 73% of male participants chose a competitive incentive scheme compared with only 35% of women. The idea that men are more competitive than women is sometimes used to justify the dominance of men in certain industries or in positions of power, but this is a bad argument.

---

18 M. Niederle and L. Vesterlund, 'Do women shy away from competition? Do men compete too much?', *Quarterly Journal of Economics* vol. 122, no. 3 (2007), pp. 1067–1101.

Competition and gender

In the framework of weak arguments I introduced in Chapter 2, there is a weak argument that looks like this if we keep thinking along gendered lines:

1. Men are observed to be more competitive than women.
2. Being competitive is better for success.
3. Therefore men are more successful than women.

And then we are in danger of being pushed into this dichotomy, as in Chapter 3:

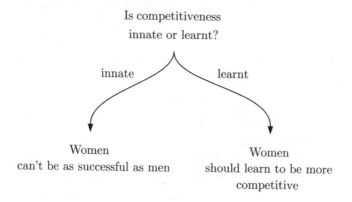

In this argument 'competitiveness' could be replaced more generally with 'ingressivity'. Once we acknowledge that ingressivity is largely better for success *because society is ingressive*, then we can think more clearly through this argument instead of being pushed into the innate/learnt dichotomy. Society favours ingressive behaviour, so it seems necessary for success. Men are currently more ingressive and so more successful. Currently women are more congressive, so they're being excluded. Moreover congressive behaviour is better for us, so the exclusion of congressive people is bad for all of us.

Favouring women is one solution but keeps us stuck in the gender dimension. Favouring congressive behaviour because it's better for us frees us from that trap. Incidentally I also think it is actively feminist because it's fighting against a status quo that favours men, and it has the potential to be less divisive because we're not fighting men specifically. Indeed, men can also succeed in the new structures by being more congressive. The difficult question is how to change the status quo when society and individuals are clinging to it.

Clinging to the status quo

We've seen various ways in which, contrary to some wide-spread assumptions, being ingressive might not be beneficial. So why do we cling to it? If we do, then we're helping to maintain the power structures of the status quo the way they are. We might say that if we're not the people in power then we can't try and change the system, and that if the people in power don't want to change it then it will never change.

However, if we believe we can't change anything, we really won't be able to. We should make sure we're not just telling ourselves we are powerless to justify our inertia.

General inertia holds us back in many ways. It is often much easier to carry on doing the same thing rather than try and change it, especially when you can't be sure what will happen if you do change it. You might not believe it's possible to alter it, or you might not believe that the result will be better, like women who benefit from white men's power so don't feel motivated to overthrow it. Then it seems to come down to the people who are most disadvantaged in the current system to do all the work of reforming it, as they are the ones with the least to lose. But they are also the ones with the least power to change anything.

An added conundrum here is that it might seem that it would take ingressivity to dare to change anything, but if it's only the congressive people who want things to be done differently then we're stuck. The ingressive people who benefit directly from the ingressive system will of course always seek to keep it the way it is. I believe this currently largely benefits men, along with women who succesfully emulate ingressive behaviour. But again it's not all men, and it's not all women, so thinking in terms of ingressive and congressive behaviour can help us become clearer about what is really going on and how to change it without just excluding men in return.

I am going to argue that change is possible, that almost everyone will be better off if we effect that change – in fact everyone except those who currently hold disproportionate power. This is about sharing power according to character traits rather than fixed identities, and so I believe it will empower women as well as any other under-represented and marginalised groups.

Crucially, I believe that we can make this change congressively. As a first step, we can start to break the cycle by severing the link between the ingressiveness of society and our own ingressive behaviour. We can find ways to operate congressively even within an ingressive society. This is the subject of the next chapter.

6

Leaning out

I eventually realised that I was unhappy with the standard ingressive academic environment, unhappy with how my character was being influenced by it, and unhappy with the effect I perceived it to be having on students. So I left the conventional academic career path and built myself a congressive new career based on sharing my love of mathematics as widely as possible, bringing more people into maths, and removing the obstacles and barriers around the world of maths. I have left behind the ingressive principles of exclusivity, evaluation, competition and 'brilliance' that I didn't like in mainstream academia. In my congressive new career it so happens that I work with a vastly higher proportion of women than I used to. There seem to be far more women in the parts of maths to do with outreach and inclusivity than there are in research departments. Of course, they're not all women – they're largely congressive people, so I get to meet congressive people of all genders who care about these principles. Many of them are in fact professors in research departments, trying individually to fight against the ingressive environment on behalf of their students and the wider population.

Perhaps if I went back into the ingressive academic world now, equipped with my understanding of ingressive and congressive behaviour, I could work out how to deal with it without becoming as ingressive as the environment. Better still, perhaps now I could work out how to change the environment so that it would become more congressive, and thus encourage everyone in it to become more congressive as well. As Ruth Whippman writes, 'Enough leaning in. Let's tell men to lean out.'[1] But, again, it's not just men: sometime's it's women who are too ingressive, especially in fields like academia where they may have felt the need to emulate such behaviour to succeed, and having done so might be rather proud of it and suspicious of any suggestion that another way is possible. But I would like everyone, men and otherwise, to become more congressive, and I believe that we can do this even inside our existing ingressive power structures, while we are still in the process of trying to change them.

At the heart of our existing power structures is a system of assumptions that is propping them up, and keeping the balance of power away from congressive people. As I've said before, there is also a system of direct prejudice keeping women and minorities out of power, but I think it's beneficial to think about these issues along separate dimensions to make sure we address them both.

The assumptions I'm talking about are that certain behaviours are important for success, and that the best way to achieve those is something ingressive. This could be self-confidence, risk-taking or resilience, for example. This diagram depicts that system of assumptions, with Y representing any of those qualities in question:

1 Ruth Whippman, 'Enough leaning in. Let's tell men to lean out', *New York Times*, 10 October 2019.

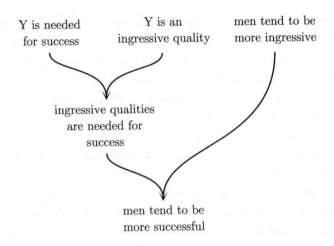

Y is needed for success

Y is an ingressive quality

men tend to be more ingressive

ingressive qualities are needed for success

men tend to be more successful

The three basic factors that seem to be causing the outcome are listed at the top. The right-hand side is gendered thinking, and addressing it is essentially the idea behind 'leaning in', that women need to become more ingressive in order to be sucessful. However, if we consider the middle and left-hand factors instead then we are moving away from directly thinking about gender differences. We might still get stuck in ingressive assumptions but we have a chance to move away from that too. I am going to challenge the middle factor, and argue that we can find congressive approaches to things that seem ingressive, such as self-confidence, and that maybe the congressive version is so different that it needs a different name. I will also address the left-hand factor, and the possibility that actually some very different, more obviously congressive qualities can be better for success even while we are in a largely ingressive world. This is how I have managed to unlearn my learnt ingressive behaviour and build a congressive career that is much more in line with my true values. I meet many women who share those values and are put off by the pressure to be more ingressive, and I believe these ideas will help them participate without going against their values. But with our new ungendered terminology we see that this will not just help women. I have decided to speak up for congressive people who do not want to emulate ingressive behaviour. I think they are losing out in society. With many types of opposite characteristics it is easiest to say that we need a balance of things. Would it be better to have a balance of ingressive and congressive traits, and encourage congressive people to become more ingressive as well as ingressive people to become more congressive?

I think it's unnecessarily divisive to ask which is 'better', and more productive simply to look at different situations and consider in what way each type of behaviour is beneficial. I see only a few situations in which ingressive behaviour is really important, and find that in most of those cases it has downsides that go with it. And that in fact the benefit could be achieved in a congressive way without the attendant disadvantages.

Anyway if, like me, you don't like being ingressive or feel you can't be, then the question is moot. It would be better if I were taller – but not too tall. Being short is largely inconvenient, except on rare occasions such as folding oneself into a plane seat or taking a nap on a sofa. However, it's a bit pointless saying it would be better to be taller, because I can't be taller. It's more productive to find ways to mitigate being short, like having a library-style rolling step stool in my kitchen so that I can reach the higher shelves.

So if we feel that we can't or don't want to be any more ingressive then it's more productive for us to think of congressive ways to deal with the world. The previous chapter was about how the structures of society tend to be ingressive, and how that nurtures ingressivity in people, which in turn makes society more ingressive. Now we're going to look at how individuals can be congressive even in an ingressive society, and how small-scale congressive interactions could gradually influence society to become more congressive as a result.

The study of how individuals interact with structures is widespread in maths, although in that case the 'individuals' are mathematical objects rather than people. We study large structures broadly and we zoom in and study the small pieces that fit together to make those large structures. We study different ways in which the small pieces could fit together to make different large structures, and we also study ways in which different small pieces could be used to build the same large structures.

One example of this is with maps. We can make flat maps of small regions of the earth such as an individual town, but if we stick them all together we need to add in some curvature in order to get a whole globe. If we stick them together too simply then we will make a flat earth, which – despite the beliefs of a few deluded people – is not correct. However, it's useful to have small flat maps of local areas because most of the time we humans are only moving around small areas of the world.

The interaction between local properties and global properties is a big part of maths because we often try to build up our understanding of situations by starting with small building blocks and then building them up gradually into larger things. Then, faced with a large and complicated structure, we can try and break it down into small parts and understand how they fit together into the whole.

One of the powerful aspects of category theory is that it works very well as a framework at zoomed-out levels and also at zoomed-in levels. We can look at relationships between whole worlds, relationships between objects within worlds, relationships between different parts of an object, and so on. This flexibility of levels is an advantage of abstract theories. In this chapter I'm going to zoom in and look at how we as individuals can operate congressively even within an ingressive society.

I offer this as an alternative to the idea of 'leaning in', for more congressive people who aren't comfortable with that idea. We have previously seen that when we are still stuck in the dimension of gender we get caught in a trap of thinking that women either have to become more like men and 'lean in' to be successful, or resign themselves to not being successful. Or we attempt to address male domination by trying to get men to be 'more like women'. Now we can rephrase the problem like this: everyone might think they have to learn to be ingressive or else give up on success. Some people are more comfortable learning ingressive behaviour than others. I did it for a while and then realised I was not comfortable with it, but this did not mean resigning myself to not being successful. There is a third way, which involves neutralising the ingressive pressures of society and finding success congressively.

First I would like to present some role models of people who have found ways to become more congressive over the course of their lives, to help convince us that it is possible and worthwhile.

Congressive role models

I found Michelle Obama's memoir *Becoming* eye-opening in many ways, but I'd like to focus on the part of her story that seems to me to be about ingressive and congressive versions of success. She writes of pursuing success first according to the usual definitions of success: going to a prestigious law school, getting a job at a prestigious law firm, earning a huge salary. She achieved those things, but then she realised that wasn't what *she* wanted for herself. Perhaps she felt she had to do it first to prove that she, a black woman from modest beginnings on the South Side of Chicago, could. But then she wanted a different kind of 'success', success evaluated on her own terms in work that she found meaningful. She asked herself, 'How do I want to contribute to the world?' and realised that the job she had as a hot-shot lawyer wasn't it.

She spent some time investigating ways in which she could make a more meaningful contribution with her expertise and qualifications and switched career, first working as an assistant to the mayor (with a 50% pay cut), then at Public Allies, an organisation recruiting promising young people who might otherwise be overlooked, and giving them training and support for non-profit or public-sector work. After that she worked on community relations for the University of Chicago to bring down the walls between the university and the South Side neighbourhood that is all around it but so often excluded from it. She calls these more 'civic-minded' jobs rather than the '*my-isn't-that-impressive* path'.

I would say that she first pursued a traditionally ingressive career as a lawyer, was successful at it, and then realised that she really wanted a more congressive career and a more congressive form of success. She congressively went about working out how to do that, thinking about her values and consulting people she respected. It is telling that she first felt she had to prove herself in the traditionally ingressive career path. At the risk of sounding like I'm trying to compare myself with Michelle Obama, I will say that I also felt I had to prove myself in the traditionally ingressive career path of academia before I felt I could re-examine my values and create a more congressive career for myself. Oddly enough, perhaps if I had been more ingressive I would have felt able to reject the traditional framework earlier.

In *The Art of War for Women*, Chin-Ning Chu says: 'We have trained our minds to think of success in a certain way – the *male* way; it's only about getting ahead, climbing the corporate ladder, becoming CEO.' I would, of course, call it the *ingressive* way, to extract ourself from gendered arguments. Extracting ourselves in this way can also help us see that we don't necessarily need female role models; we do need them to help us see beyond gendered oppression, but to help us see beyond ingressive assumptions we need congressive role models of any gender.

When looking for such role models we can run into a vicious cycle: if society celebrates ingressivity then we will hear about ingressive people more. But other stories are there if we look for them.

Prof. John Baez is a physicist and prolific blogger who started sharing his understanding of physics way back in 1993 when the internet was barely known. He wrote a weekly column, 'This Week's Finds in Mathematical Physics', which was a congressive way of helping the world and himself at the same time: he knew that the best way for him to understand things for himself was to write about them for other people. Many great teachers find this about teaching, that you understand something much more deeply when you're motivated by the desire to help others understand it. This is one way in which congressivity helps oneself. It is very different from the kind of teaching which consists of posturing at the front of a lecture room and relishing the power you have over the hapless students. Baez is so focused on helping society that he thought about what he could best do with his mathematical expertise to help the world, and decided this meant shifting into working to save us from the global ecological crisis. John inspired me to think similarly deeply myself, about how I could best use my mathematical expertise to help the world. Although my answer is quite different in specifics, it is the same idea.

When I began my career my ambition was to be a professor of mathematics at a top university, leading a powerful research group and attracting brilliant PhD students. Now my ambition is to change the world to bring more people *into* things, people who have previously been excluded: from mathematics, from science, from education, from full participation in society. Remember, it doesn't have to be a competition, but I might actually say my new ambition is *more* ambitious than my old one, although I think it is a congressive ambition whereas the old one was ingressive. In making these changes to my career and life I was greatly inspired by John Baez.

I have needed role models who are women, to help me believe that it is possible to be a successful woman without emulating male behaviour, but it is helpful for me to see congressive role models regardless of their gender, such as Baez. It's particularly important to separate those issues out because so many successful women who are held up as role models are in fact rather ingressive. Congressive female role models would be most useful, but congressive male role models are much more helpful to congressive people than ingressive female ones.

Exceptions to this are ingressive women who have consciously worked hard to become more congressive, for the good of people around them. Professor Emily Riehl is an abstract mathematician currently working at Johns Hopkins University. She also happens to have competed in Australian football at the international level, representing the US, and by her own admission is extremely competitive and ingressive. This stood her in good stead when she was applying for grad school, making her way as a young researcher, getting her research known, and introducing herself to senior professors at conferences.

But as she became more senior she realised that a more congressive approach was important when she was teaching and nurturing new students. She runs many seminars and workshops for grad students and events for women and minorities, and stresses at the beginning that, essentially, if she is being too ingressive about her teaching then she really wants people to ask questions and get her to be more congressive. I recently asked her at a conference how she manages to ask so many questions during people's talks, which I still don't dare do in front of an audience in case my questions are considered stupid. It turns out that it might seem ingressive of her but she's really doing it congressively – she knows that it helps other people and it helps to change the tone of a talk, so she has committed herself to ask a question at every single talk she goes to. She admitted that she was terrified of doing it at first but she forced herself to anyway. One might call that an ingressive approach to being congressive – forcing oneself to do something unpalatable to help others.

It might seem easier to dare to change if you are more obviously ingressive, but personally I find that helping others is a very strong congressive motivation for doing things that seem daring. In fact, this is a specific example of a general principle: we can look for congressive ways of doing things that seem on the face of it to be ingressive, like daring to change.

This might involve reframing many things we take for granted, in further acts of digging down to first principles. I have spent the last few years doing this in my life. I have found that becoming more congressive has not just been a subsitute way to motivate myself to do ingressive-looking things, but further it has opened up far greater potential in me, and has enabled me to make better contributions to society than when I was trying to be ingressive. This is aside from it simply being abstractly eye-opening to see the extent to which we make unnecessary ingressive assumptions about life. Also I am happier.

In the two-dimensional plane of ingressivity and congressivity I think I've broadly moved over the course of my life somewhat like this:

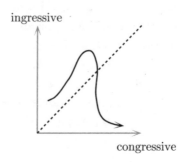

The gradual rise shows that as I learnt to be more ingressive (competitive, combative, ambitious) I also learnt to be more congressive (collaborative, nurturing, actively helping and bringing people together). The peak shows the height of my attempt to respond to society's rewards by trying to be 'successful' in a standard academic career. Then, when I started unlearning the ingressive behaviour, I didn't yet know how to be actively more congressive, and I'm still learning that.

Some people might be equally ingressive and congressive by being neither very actively, so they'd be near the bottom left. Others might be very actively both things in different situations (or perhaps at the same time) and they'd be up at the top right. It's important to remember that in this two-dimensional version of a spectrum there are many different ways to be 'equally' both things – anywhere along the dotted diagonal.

A full representation of the gradually shifting nuances of the two-dimensional spectrum is only really possible if we use two colours; this is the red and blue square incorporated in my author photo.[2] The two colours can gradually get darker as we move up or right, and they can also gradually blend together towards the middle.

As it's not a clean dichotomy the shift to more congressive behaviour can happen gradually by tiny increments. It doesn't have to be a sudden flip. It can also happen in some situations but not others, just as I became more congressive in non-work situations before I was able to become more congressive in work situations.

---

2 If you're not reading a physical book with my author photo on the jacket, you can view it at eugeniacheng.com/square.

It might sound impossible to be congressive in an ingressive world without changing the structures first, and it might also sound impossible to change the structures, in which case are we stuck? What I do know is that if we assume it's impossible to change anything then it will be impossible. I also know that small changes are possible, and that they can build up into big ones. So I'll start with small changes we can make to our own personal ways of thinking.

Reframing ingressive assumptions

As we're very used to our ingressive society we may well be thinking ingressively without realising it. It may seem that we have to be ingressive to succeed, but instead we can reframe some ingressive-sounding behaviour so that it becomes congressive. We may have been looking at things from one perspective only, given that through most of history successful people have been men. This gives us a one-dimensional view of what success looks like, and has resulted in women emulating men's behaviour in order to try and achieve equality with them.

In fact it might require even more ingressivity for women to emulate the behaviour of men, because men do actually experience the world differently. It might not take much daring for a white man to speak up in a group setting because he suffers no social threat, and may feel no negative effects if he turns out to be wrong. Whereas it might take much more confidence for a woman to speak up in the same situation if she is in genuine danger of being ridiculed, as women often are. She may also feel more hurt by ridicule than the man does, so even if people do laugh at the man for being 'stupid' he might find it easier to brush it off. This is all to say that if we emulate men's behaviour we are not necessarily having the same experiences in the process.

Instead of ingressively facing such 'risk' head-on, we can do something congressive about it. In the end I think it would be much better if everyone learnt not to mock other people, but in the meantime we can think of ways to handle such situations congressively rather than just ploughing through. I am aware that there are many ways in which I might look ingressive, but my motivations are actually congressive. I'm really not inherently self-confident, but I stand up and speak in front of large audiences. I avoid taking risks but I took the apparently very risky step of quitting a secure, permanent job to become largely freelance. I am easily hurt, and yet I put myself on social media at the mercy of all the random obnoxious men on there (and yes, the people who are obnoxious to me on social media are almost always men). This is because I have found congressive ways to do things that look ingressive.

Understanding this possibility of reframing can especially help people who feel put off by ingressive situations, as happens with many women. It can also help us change the atmosphere in any situation to help congressive people participate. I think this is an important aspect of inclusivity that will in particular help gender imbalance.

Take the example of 'strength'. It might initially seem that strength is more ingressive than congressive, but there are different ways to be strong. Traditional strength might be thought of as power, toughness, the ability to be unyielding either physically or emotionally, and thus overcome others. That is ingressive strength.

Congressive strength could be more like flexibility. Some physical materials are strong precisely because they are flexible, whereas something rigid is more likely to break. Congressive strength is like a river altering its course to wind its way through a contoured landscape, as opposed to a waterfall that just pours out. Or deep-rooted plants that can sway in strong winds, compared with a big tree that is more likely to fall down.

Another congressive form of strength is to know when and how to get help, and to draw support from people around you. You don't have to be independent and self-reliant in order to be successful. Our ingressive society celebrates independence and 'self-made millionaires', but nobody is really self-made. This is one way in which our gendered society can be detrimental to men, as getting help is seen as a sign of weakness in them. And thus many men, through society's misguided pressures, miss out on the benefits of support, and even medical care.

Nobody accomplishes anything in isolation from the people around them and society at large. Building a good network of support is a congressive alternative to self-reliance, a way of building interdependence rather than independence. (It's also how I deal with the stress of obnoxious people on social media.) It can be more efficient and even fun, too, like taking it in turns to cook for a group of people rather than everyone cooking for themselves alone.

Related to self-reliance are self-esteem and self-confidence. Those might sound like ingressive traits, but I think it depends where the self-belief comes from. Let us go a bit further into first principles and think about what self-esteem means. It can sound like it is belief in yourself that comes from nowhere but yourself, a bit like being self-made or self-reliant. That would indeed be ingressive, and I've realised that those ingressive undertones put me off: I feel I don't want that sort of self-esteem.

But belief in yourself can also come from external validation or preparation. If you achieve some things in the world, then it doesn't take self-reliance to start believing in yourself. Likewise if you have prepared very thoroughly for something. If you have achieved nothing or are woefully unprepared but still for some reason believe in yourself, then I would call that ingressive. Those are two different types of self-esteem. Here is a diagram depicting these different ways of having or lacking self-confidence, relative to external validation or preparation:

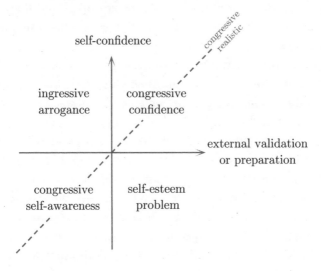

I definitely feel that the confidence I have comes from external validation and preparation. I dreamt of being a maths professor but wasn't sure if I could do it. So I worked hard, and every time I passed another hurdle I became more confident. But I still made long lists of back-up plans in case I didn't make it. My confidence grew as more senior professors encouraged me, and I feel I gradually inched my way into the top right. By contrast, ingressive aspiring mathematicians keep believing they can do it even when rejected repeatedly; they are in the top left. Needing external validation isn't necessarily a sign of insecurity; it could be about making sure you're realistic rather than arrogant or delusional.

On the bottom half of the diagram are different ways of lacking self-confidence: on the right, if you have plenty of external validation and are very well prepared but still don't believe in yourself, then I'd say you might have a self-esteem problem (or stage fright). Being congressive doesn't mean refusing to believe in yourself even when everyone else does.

The bottom left is where you have plenty of external evidence that you're not good at something, or you know you're woefully underprepared, in which case it's self-aware and realistic to acknowledge it. It also might be more productive. I have plenty of evidence that I'm not good at finding my way by instinct, so I always follow a map; I know some other people who always get lost but *still* don't use a map.

The dashed diagonal line shows a sort of 'balanced' congressivity: a view of oneself that is carefully rooted in external validation. Someone might argue that it's productive to be a *little* more ingressive than that, so that you have some intrinsic sense of your own worth regardless of the unpredictable nature of the world's outcomes. That might be true, but then I'd also argue that I'd rather this were instilled in childhood by the right sort of validation from parents and educators, in which case that apparently intrinsic self-worth would actually be based in external validation as well. And understanding that things like exam results and prizes can be a bit unpredictable (and contentious) is, I think, part of having a balanced response to them.

In yet further nuance, there are ingressive and congressive forms of validation too. Ingressive validation is in the form of awards, prizes, grades, income, status symbols and other external signs of your 'superiority' to others. Congressive validation is in the knowledge that you have made some sort of contribution, helped some people, created something, improved something.

A particularly pernicious form of ingressive validation is if you bolster your feeling of superiority by putting other people down. This can be in the form of trying to make them feel inferior because of their knowledge, tastes, job or social standing, or excluding them from something so that you feel special. Prejudice and bigotry are ingressive ways to bolster your self-esteem by keeping other people out of something. They involve declaring some other group inferior in order to declare your own group to be superior. In *Testosterone Rex* Cordelia Fine writes of a study finding that men's self-esteem went up when they were told they had done badly at something at which women tend to excel, because 'incompetence in low-status femininity helps establish high-status manliness'.

Arguably, that type of need for feelings of superiority stems from deep insecurity – the opposite of self-esteem. Perhaps in that case it actually takes more self-esteem to be congressive and not need that feeling of superiority. And of course, instead of associating character traits with gender and thus considering male ones to be superior, we can now look at them without associating them with gender and evaluate them for what they are.

Self-confidence that comes from external validation could be thought of as a congressive form of self-confidence, but the very concept of 'self-confidence' sounds ingressive and might put congressive people off; it definitely put me off in the past. There are many other character traits widely considered important to success that similarly sound ingressive, but we can look for congressive versions of all of them. Another one that I've mentioned with respect to me quitting my job is risk-taking.

Risk-taking

Our ingressive society reveres and rewards risk-taking. At a superficial level we marvel at people who do things like climb up rock faces with just their bare hands, or tightrope-walk between skyscrapers. At a less superficial level, high-risk research in science is often seen as more exciting. This might take the form of implausible experiments that show something quite unbelievable: it's not that there's a risk to anyone's safety, there's just a high risk of failure. We'll come back to ways in which low-risk research can be more valuable, and how to nurture it, in the next chapter. Risk and gender inequality are linked in subtle ways and a great deal has been written on that topic elsewhere (see *Testosterone Rex*, for example). It has been shown that women exhibit less risk-taking behaviour in general, but that this is not an innate biological difference. The aspect I want to address is the way in which women are then exhorted to learn to take more risks in order to be as successful as men: it's yet another form of 'leaning in'.

I'd like to give a sort of blueprint for how we can learn to 'lean out' instead, that is, reframe apparently ingressive behaviours in a congressive way, and then take congressive action instead. This involves going back to first principles (again) and asking ourselves what it really means to 'take risk'.

Here is a diagram depicting some different responses to risk.

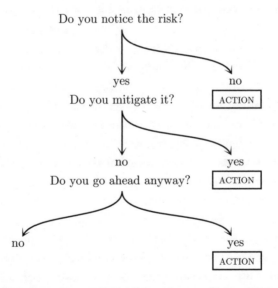

The left-hand side is where the risk actually prevents you from taking action, whereas any of the outcomes on the right-hand side (marked 'ACTION') could be described as 'taking a risk'. However, I think they have very different flavours. In the middle version you mitigate the risk, perhaps by building a huge safety net, in which case you're not really taking a risk at all as all the risk has gone. I don't think this version is what is generally meant by 'risk-taking', but in particular it's not what I *feel* when ingressive people insist that you must take risks to be successful; I think congressive people might rather feel they're being asked to do the top or bottom versions, where you either block out the risk or charge into it.

It's also important to remember that the same situation can involve genuinely different amounts of risk for different people, as we already discussed in the context of people speaking up in group situations, where there is genuinely more risk to women than to men. Big strong men really are at less risk than me walking up the street at night. More abstractly, people with high social status can take bigger social risks, or rather, it's not actually a risk.

Of course, ideally we would remove the risk by stopping people doing that belittling or other obstructively ingressive behaviour, but in the meantime what can congressive people do? If congressive people feel that they must take risks to be successful, they might be put off even trying. Instead, I urge congressive people to build big safety nets to reduce the risk. This can be in the form of contingency plans, careful preparation, or a support network. And this doesn't just have to be defensive behaviour, it can be actively productive too, and thus something that more ingressive people could benefit from learning. I am an extremely risk-averse person, but rather than hampering me this has caused me to become much better at things through my efforts at reducing those risks. I honestly feel that I have not taken substantial risks in my life.

I invite you to question any received wisdom or advice about how to be successful, and ask whether it is making ingressive assumptions. If so, and if you or anyone else finds that offputting, you could try reframing the advice in a congressive way or finding a congressive way to approach the same sort of outcome.

An analysis along these lines can help us find congressive approaches to other ingressive-sounding attitudes or behaviours that ingressive society tells us we should learn, such as 'leaving your comfort zone' or 'resilience'. Instead of leaving my comfort zone (which sounds terrifying), I work out how to extend my comfort zone. Resilience is also an offputting idea to me, because I don't like the idea of bouncing back, or of being somehow impervious to bad things happening. I would rather be a sensitive human who is hurt by bad things, but then I can try and find ways to transform the bad experiences into something good, such as a way in which I can help other people.

For some, this is a form of resilience, but if the whole notion of resilience puts people off, then I think we should extract the part of the concept that isn't offputting and call it something else, as with 'congressive risk-taking', which was really about building safety nets.

First, as with risk-taking, I want to stress that resilience or the appearance of it is much easier for privileged people. You can always weather misfortune better if you have more money (all other things being equal), or more social status (by being white, or male, or straight, or cisgendered), or better health. We should be careful about whether asking people to become more resilient in order to be successful is insensitive to their disadvantages in life.

But I also worry that insisting on resilience can be used as a form of oppression, masquerading as a way to help marginalised people progress. Women are told that they need to be more resilient in the workplace, but this often means that they should just put up with bad behaviour, bigotry, sexist jokes and harassment. If I had been more resilient I would have stayed much longer in a job I didn't like, thinking I should just put up with feeling undervalued, and bullied in ways that might have been racist, sexist or ageist.

Being congressive in an ingressive environment does not mean putting up with sexism instead of fighting back; it's about learning to thrive, if possible, or working out how to get out if it's not possible to thrive. Perhaps 'congressive resilience' is really about transformation. Rather than put up with an ingressive situation we could try to transform it.

Congressive responses to ingressive energy

How can we deal with an ingressive situation without either being overwhelmed by it or becoming ingressive in turn? How can congressive people make themselves heard in the face of incoming ingressive energy? I previously trained myself to be ingressive right back, but every time I did it I would go away and really dislike myself. And it didn't actually achieve anything except perpetuate and escalate the ingressive energy in the world.

But that seemed like the only option other than just letting the ingressive people walk all over me, which might not escalate the ingressivity but certainly enabled it. Congressive people often get walked on because we're trying so hard not to be horrible people. In particular, women are pressurised by society to sacrifice themselves for the good of other people, either in their families or in caring professions. It is seen as noble for a woman to put others first even to the point of erasing herself. I have too often been accused of being selfish when I put up boundaries to try and protect myself.

However, having boundaries doesn't make you a horrible person. Protecting yourself doesn't make you a horrible person. If protecting yourself means hurting someone else, that means the relationship is a toxic, zero-sum one. Usually the other person cultivated it like that in order to emotionally blackmail the more congressive person into doing something they would rather not do.

Psychologist Philippa Perry talks about bringing up children in a non-judgemental environment rather than in one that's all about winning and losing, which enables them better to deal with combative energy by thinking, 'Hmmm, interesting person,' rather than fighting back.[3] The idea of neutralising combative energy reminds me of the Japanese martial art Aikido, which, as I understand it, involves neutralising incoming aggressive energy (perhaps into the ground) and opening up a non-violent space for interaction instead. (I am not an expert or any kind of practitioner.) The de-escalation of physical violence would be a particular valuable form of this, but de-escalation of general ingressive behaviour is valuable too. This is also a way to encourage congressive behaviour in others and a starting point for building new congressive structures. It does take practice,

---

3 Robyn Wilder, 'Philippa Perry: "Listen carefully, parents – and don't despair"', *Guardian*, 10 March 2019.

and doesn't always work, but that can be reframed congressively as well.

Here is a small example that I have worked on for a few years. It involves someone insisting they want to discuss something with me on the phone. This might sound just normal. Indeed, some popular wisdom says that it is much better or more efficient to discuss things on the phone. However, I think this assumption deserves to be challenged.

I have found that sometimes people want to use the phone just because they don't want to take the time to organise their thoughts into a written form. But sometimes it's because they want to put me on the spot and pressurise me into doing something I don't really want to do, like more work or accept less pay or cause some other inconvenience to myself. This is some ingressive energy that I want to neutralise. I can do it quite simply: by using email instead of a phone call. Really ingressive people can get very upset about this and I sometimes wonder if it's because they know (consciously or unconsciously) that I have levelled the playing field on which they hoped to have an unfair ingressive advantage. I took my inspiration from a wonderfully congressive vice-principal of a school, who commissioned some work from me but specifically suggested discussing terms by email, observing that it can be very uncomfortable doing so in person. It is often said that women don't like asking for pay rises, but I think that congressive people of any gender probably don't like asking for a raise, or for anything at all. I think one way to make that act require less ingressivity from me is to do it by email instead of having to do it in person.

Every time I feel ingressive energy coming at me (which is often) I take it as an opportunity to practise this process of neutralising it congressively. Here is a picture of how I view the options when faced with incoming ingressive energy:

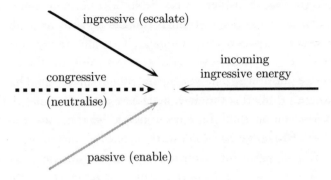

I find it good practice to try and think up three types of response, as in the diagram above. I also find it instructive to imagine what is the most immediate reaction that would pop into my head, and also which one I would previously have fantasised about when trying to come up with acerbic put-downs. In the Appendix I've given some real examples of ingressive things that people have said to me, and three types of response I have thought of.

I have found that this is a subtle way to shift the balance of power in an interaction with someone behaving ingressively. As men currently hold more power in society, structurally, it is unfortunately typically men who try to wield ingressive power over me. But I have found that they respond much better to my congressive neutralising of that power than if I try and wield ingressive power back.

It is hard, because of years spent jumping to ingressive responses, but there is no risk involved if you frame it as having no chance of failure: if your aim is to neutralise the ingressive energy then it is possible you will fail, but if you reframe this congressively as aiming to *get some practice* at neutralising the ingressive energy and build up some more case studies, then there is not really a concept of failure. No matter what the outcome is you can go away and think up different types of response later. Typically we all ingressively go away afterwards and try and think up the best acerbic put-downs that we can, and then wish we could have thought of them at the time. But I have come to realise that I would rather think of ways to neutralise the ingressivity rather than come up with a scathing retort.

When we neutralise ingressive energy we achieve several things. We open up a space for more congressive interaction, but we also move away from rewarding ingressive behaviour and towards encouraging congressive behaviour. And, crucially, we change the balance of power away from the status quo. As the status quo currently favours men, I consider every act that tries to shift that power to be an act of feminism, no matter how small it is, and no matter how well it seems to go. Just as individuals can be contributing to the oppression of women even if they're not specifically misogynistic, we can also contribute to addressing gender inequality even if we're not specifically helping women. I truly believe this is the power of congressivity.

This is a good start towards our ultimate aim, which is bigger than just learning to be more congressive ourselves and encouraging congressive behaviour in others: it is to create a more congressive and thus more inclusive world. This is the subject of the next and last chapter.

7

Dreams for the future

Here is my dream train design:

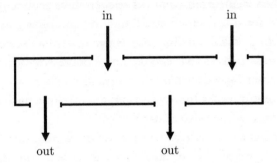

Every train platform would use both sides of the train: one side for entering and the other side for leaving. This solves the problem of people crowding round the door and not letting others off before trying to get on. The offset of the doors on either side solves the problem of people not moving down to the middle of the carriage because they're so determined to be near the door in order to get off. This is a tiny microcosm of a structure that has congressiveness built into it so that the advantage of ingressive behaviour is neutralised. This particular example is mostly a fantasy as it would be ludicrously expensive to rebuild all existing train stations and trains to this design. Even so, situations with entrance and exit on opposite sides do exist on a small scale, such as on light rails at airports for transfer between terminals, in some lifts, and often at underground train stations. I admit to feeling slightly gleeful when I get in a lift where I know the exit is going to be on the other side, and I watch inconsiderate ingressive people crowd around by the entrance without realising that not only are they inconveniencing everyone else, but they're inconveniencing themselves as well.

I'm going to propose some dreams for the future, ways we could restructure the whole of society to be more congressive. In some cases I will build from situations that already have congressive possibilities, or perhaps small-scale congressive structures that could become more widespread. In other cases I will dream freely to help us break away from of our assumptions, even if, as with the trains, practical implementation is harder to envisage.

This is my answer to all that is written about feminism that says we need to change the entire system, but doesn't suggest how we could do that. I think we can address this, equipped with our new ideas from the second part of the book. In particular I think we can tackle the problems we described in the first part that were to do with still being bogged down in one-dimensional thinking along lines of gender. We were stuck thinking along the lines of men and women being 'different', and being distracted by questions of whether those difference are innate or not. So we were obliged to make a false choice between various gendered ideas as to how to change the status quo.

We can now move away from all the gender-based choices. We can move away from pseudo-feminism or 'leaning in', in which women are exhorted to become more like men in order to be successful. We can also move away from the opposite, in which men are asked to become more like women. We can move away from 'reverse sexism', in which women are deliberately favoured to make up for past oppression, and away from anti-feminism, in which women are told they simply biologically don't have the characteristics to be successful.

Instead, with the ideas of ingressive and congressive behaviour, we can free ourselves from that one-dimensional trap and consider how to make more congressive structures in society, which will stop us favouring counterproductive ingressive behaviour. I believe that this in turn will bring out the even greater potential of more congressive people and encourage everyone towards greater congressivity, which will in turn strengthen the congressive aspects of our new structures.

Many mathematical advances come from making tiny increments that build together into enormous wholes. To try and understand or change the whole structure at once is indeed daunting, but if we understand and change small parts and build those small parts up into large ones I believe that we can make a difference, as I have already done in various aspects of my life at many levels, including personal, interpersonal, and professional. Pure maths is a place of dreams. It's about dreaming up new concepts and new structures. Sometimes it's about dreaming up new concepts that generate new structures, like with imaginary numbers. Sometimes it's about dreaming up new structures for old concepts, like category theory.

Unfortunately, maths too often seems like a place of

rules, a place of rigidity and constraint. While that is true at some level, there is a deeper level that becomes increasingly important as you get further into research. The deeper level is about understanding the *point* of the rules and constraints. If you understand exactly what the nature of the constraints is, then you can move more freely. If you understand exactly what the rules were there for, then you can dream up productive ways to break those rules and create better worlds.

In the previous chapter we discussed ways in which individuals could become more congressive, even within ingressive systems. But if we can collaborate to make more congressive systems, then it won't matter that some people are still very ingressive as that energy will be dissipated. And I think we can make a virtuous circle, as individuals influence the structures and the structures in turn influence more individuals.

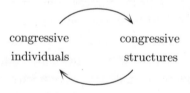

congressive   congressive
individuals   structures

This is a step further than just neutralising ingressive behaviour.

One abstract example of this is the 'prisoner's dilemma', which I (and many others) have written about elsewhere. It's a favourite thought experiment in philosophy and game theory showing that collaboration can give better results than competition. In this hypothetical situation two prisoners are given the opportunity to denounce each other. If neither of them does, then they will be sentenced to, say, 2 years each. If they both denounce each other they will get 5 years each, but if just one of them denounces the other the denouncer will go free and the person who stays loyal will get 10 years.

If the prisoners are not allowed to collaborate with each other this is tricky. If the other person denounces you it would be much better also to have denounced them, so that you get 5 years instead of 10. If the other person doesn't denounce you then it would still be better to have denounced them, because you'll go free instead of getting 2 years. So if each person acts as an individual they will denounce each other and they will each get 5 years, and this would count as 'logical' behaviour. But it is ingressive logic. Thinking congressively they could both stay loyal and end up with a better outcome: only 2 years each. This could work even if they're not allowed to communicate with each other to collaborate, as long as they both think congressively and trust the other to think congressively as well.

This structure still relies on trust, but I like dreaming up structures that automatically induce congressive behaviour even from people who might resist it, as with my dream train design. I believe that this is a powerfully congressive way in which we can start to change the power structures of the status quo. And since those power structures currently still favour men, I think that this is a form of feminism. This might sound gendered, when the whole point of this book is to argue for an ungendered approach, but to me this is an ungendered approach to feminism.

Just as there is indirect bias built into the system even when it doesn't explicitly discriminate against women, I think we can reverse that without explicitly working along gendered lines either. This really is using our new dimension to escape the traps of our old one.

I will start with areas in which I have the most experience – maths, in terms of teaching and research, and education more broadly. But the dreams can expand and be as big as anyone wants.

Congressive maths

As we've discussed, maths is still a very problematic area when it comes to gender bias. We need to work to overcome explicit bias against girls and women in the form of prejudice, stereotype threat, and the lack of role models, but we also need to work to overcome indirect bias that might be happening ingressively through the way maths is currently being taught at all levels. A more congressive approach directly addresses the latter issue, but I think it could address explicit prejudice as well. If we focus on the value of congressive aspects of maths, we might notice that the types of people we have previously associated with mathematical brilliance (for example, men who talk themselves up a lot) are really being overvalued.

Congressive maths is what I have been teaching to art students for several years. I think it's also what is introduced to children at the very beginning of school, when it's all about play and exploration with blocks and toys and other things they can touch and feel. It comes back round to being like this at a research level, but by that time I think we've already put off far too many congressive people by the phase in between.

Dreams for the future

I think maths should be congressive all the way in between as well. We really don't need to train people to be human calculators any more, now that we have actual calculators with us more or less all the time (for example on our phones). So maths could be more congressive by being about exploration and processes. It could be more about ways of thinking than about knowledge.

I would like to see a non-cumulative curriculum so that each stage doesn't depend on the previous stage. The traditional model is more like a series of hurdles that get higher and higher and are specially designed to weed people out at each level. Not only is this ingressive, it's also counterproductive as we are not weeding out the right people.

I would like to see category theory taught at school as it has great potential for being maths as a 'way of thinking' rather than as an exercise in getting the right answer. But congressive maths is not just about the subjects we teach, it's also about the way we teach. It should be more about depth than speed, about invention and growth rather than finding the 'maths people' and separating them from the 'non-maths people'.

Some people might worry that if we move to a congressive way of teaching maths it will be more inclusive but with the side effect of lowering standards. Of course, this criticism is often levelled at the idea of getting more women into maths – some people believe that women really are worse at maths and so we'd just have to lower the standards. But I think the example from Finland shows that this does not have to be the case.

Moreover, the 'standards' in question are also ingressive standards. If we focus on what we can get people to achieve, usually as tested by some exams, then we are aiming for something dangerously limited that is unlikely to have a lasting effect on students' lives. A more congressive approach is to focus on students' experiences of maths and make sure that whatever happens they maintain their interest, curiosity and appreciation for it. If we train them to be able to spew out facts in an exam but hate maths, then as soon as the exam is over they will purge all that they have learnt from their brain and feel negatively towards maths forever, in which case I don't think we have achieved much. In fact we've achieved something negative; we'd have done better to teach no maths at all.

Whereas if we aim to nurture and strengthen interest and appreciation, then we are more likely to nurture adults who appreciate maths rather than spurn it. We need more focus on this congressive aim. A gratifyingly common response from audience members when I give a public talk or interview is: 'I wish you'd been my maths teacher at school.' It is almost always women who tell me this. The people who boast to me about their mathematical knowledge are almost always men. But again, it's not all men and it's not all women. To me this shows that the ungendered terminology can not only help us escape those divisive sweeping statements, but also address the gender imbalances that currently exist.

Moreover, I also think it will be better for research. Ingressive maths may raise the standards for ingressive exams, but not for research, which is much more congressive.

Congressive research

The world of academic research has some very ingressive aspects to it, as we have already discussed, and is even more male-dominated that earlier levels, especially in maths and science. But research is a deeply congressive activity. Using ingressive filters to select people for research is thus misguided, and if we nurture congressive students then I think we can find far more research potential in everyone. And if women have previously been disproportionately put off by the ingressive environment, this approach will also address the gender imbalance. Furthermore, if that really is true about women, then we *cannot* effectively address the gender imbalance without addressing the ingressivity. Trying to do so would result in diversity without inclusion, miserable women (like me in my academic job), and the unfortunate possibility of those unhappy women not being able to work well in the environment, thereby appearing to confirm the prejudiced view that women are not cut out for an academic career.

So I hope that the research *environment* can also become more congressive. It has been moving in that direction in a promising way, especially with the help of technology. Some very heartening congressive aspects, at least in maths, are the rise of global collaboration and the free sharing of both research and exposition, as we discussed in Chapter 5. This is better than the old-style ingressive research based on secretive competition, a race to be first, and publication in expensive 'prestigious' journals.

The system for publishing research is an example of a structure that can either push people to become more ingressive or nurture people to be more congressive, depending on how it is set up. An example of a congressive version is the 'registered reports' format[1] for publishing research. Unlike the ingressive system, where dramatic positive results are favoured, in this congressive system peer review is conducted on the research question and the methodology *before* the data is collected. Thus it is the process that is reviewed, not the answer. The idea is that if the question is interesting and the methodology is sound then the result of the experiment is interesting whether it turns out to be positive or negative. Psychology researcher Alexander Danvers writes that when this method is used, 'there are no research failures'.[2] He argues that the (ingressive) approach of trying to get 'eye-catching, wild' results does not do as much for cumulative scientific understanding, and that celebrating such high-risk work when it does go well skews scientific culture so that scientists spend more time on questions that are only of interest when the results are positive. The ingressive approach also leads to publishing results with low statistical significance, and no explanation, and to publication bias where 'failed' experiments are not reported because only 'successful' ones are of interest.

Actually, my radically congressive vision for a better research world would abolish the whole idea of publishing in journals. It might seem that peer review for journal publication is democratic and thus congressive, but it's a lot more ingressive than it sounds. Peer review is too often more like gatekeeper review than democracy, more about exclusivity

---

1 'Registered Reports: Peer review before results are known to align scientific values and practices', The Center for Open Science.
2 Alexander Danvers, 'Addressing the academic arms race', *Psychology Today*, 30 August 2019.

than about dissemination. I would rather lessen the power of the gatekeepers. This could include sharing all research freely and having a democratic review process to evaluate it on a scale rather than the all-or-nothing judgement of publication or rejection. We could also share grant money widely among more people rather than making everyone compete for a very small number of huge grants. We could allow prizes to be shared between more people, and reduce the hold of 'superstar' culture and the adulation of supposed 'geniuses'; instead we could congressively recognise that many people do many different types of helpful research, rather than celebrate a few and ignore everyone else.

The fixation on exclusivity is at the root of many of these problems. It is the congressive idea that keeping people out is a sign of quality, and that letting people in is a sign of weakness. It is related to the false dichotomy between research and teaching, in which people think breaking new ground is more important than making it possible for people to come with you, and experts who get their self-esteem from being able to do things that others can't do. A congressive research world would value teaching and also public communication. Congressive experts get self-esteem from helping others understand what they do, and remember to teach not just proficiency and expertise but love of the subject, love of learning, and love of curiosity.

Congressive education

I have already described how I have made a congressive classroom where everyone is contributing to the collective learning and ingressive behaviour provides no advantage. I am lucky to teach in a very congressive institution, an art school. Perhaps that is a more obvious environment in which to be congressive, as art is more obviously congressive than other subjects, but I think we can take inspiration from it to see how the education system at large could become more congressive.

The ingressive system seems to be driven by the idea of assessment, so perhaps we need to start by dreaming of a congressive system to replace that. Art is obviously much less about right and wrong answers and assessment is likely to be a portfolio rather than an exam. This has fewer pitfalls than if we tried to have portfolios for, say, mathematics or an essay subject. It is harder to pay someone online to produce your art for you unless it's digital art, but it's rather easy to pay someone to write your essays for you. There are even online services where you can pay maths PhD students in distant countries to do your homework for you.

There are some more congressive systems of assessment by 'standards' or 'descriptors' rather than by grades, where instead of a single ranking there is a description of various important aspects of skill in certain areas, and the aim is to achieve some proficiency in each one. This encourages us think about what skills education is really about and aim to teach proficiency in those instead of achieving certain grades.

Those skills might include abstract ones like making, following and analysing logical arguments; expressing thoughts clearly and compellingly in different forms; organising, assessing and assimilating new information; making connections between situations and providing a range of different approaches and points of view.

Then there might be more concrete life skills, such as languages, writing and communicating aimed at different audiences, presenting in public, computer programming, cooking (and other DIY and craft skills), budgeting and investing, fact-checking and debunking news stories, and awareness of history and other cultures.

I would add some important human skills, such as kindness, generosity, self-awareness, empathy, listening and validating, and not breaking people's hearts. Admittedly that list reflects my desire for more congressivity. It's possible that there are some aspects of this that exist in very progressive schools or progressive education systems, perhaps in Finland. There's also a list of my 'pet peeves' of human interaction that I sometimes wish were addressed in school: how to speak at a volume so that the people you're talking to can hear you, and nobody else can (which includes learning both to project and also lower your voice); how to avoid hitting 'reply all' to a group email; how to be on time for appointments and meet deadlines; how to walk through a door and make sure you don't let it close on the person behind you.

Those are, of course, big dreams that can't be changed overnight. However, one small area that is part of education but can be directly and quickly influenced is any sort of question time.

Congressive question time

One specific area which I have developed in a congressive way is question time after a talk. Typically questions are solicited by asking people to raise their hands, and then the questions are taken on a first-come, first-served basis. Not only is the selection process ingressive, but the whole situation of getting people to ask questions out loud in front of an audience is extremely ingressive. It does not invite congressive questions, which would be genuinely curious, but rather, it encourages ingressive 'performance'-style questions, from ingressive people who have no fear, and often whose main aim is to try and demonstrate what they perceive as their superiority to the speaker and everyone in the room. Typically the much more interesting exploratory questions come from congressive people who ask quietly afterwards rather than in front of everyone.

Eventually I devised a way to run question time that is congressive, neutralising the ingressive behaviour without falling into the one-dimensional traps. Before then a typical question time would consist of questions only from white men. In case you're wondering if I'm suffering from confirmation bias: on one occasion, at the Auckland Writers' Festival Waituhi o Tāmaki, people with questions queued up at a microphone, and after a few questions a white man shouted from the audience, 'Why are all the questions from white men?'

I have tried addressing this by asking for the first question to be from a woman, as some studies have shown that this results in more women following suit in the rest of the session. But all that happened when I tried this was that one man got so angry he then took the microphone and shouted at me for minutes on end. (The microphone was eventually taken away but he continued to yell.)

That is an example of 'reverse sexism' in the small microcosm of a question time. (Although I stress that reverse sexism doesn't actually exist according to the contemporary definition of sexism as only being about the group in power being prejudiced against the group not holding power.) It is less divisive to degender our approach and think of a congressive version of question time instead using this new dimension. Some proposals include having people write questions on post-its or on an online forum, but in my experience this is like antibiotics that kill off the standard bug and leave the most resistant strains of superbugs behind: the mildly ingressive people are neutralised, but the really ingressive people are actually emboldened by the anonymity of the situation, and the congressive aspect of human interaction is removed as I can't look the asker in the eye or understand anything instinctively about why they are asking such a question. It leaves the ingressive energy bouncing around the room.

Now, instead of asking for raised hands, I invite everyone to have discussions with those around them. I walk round the room hearing what people would like to explore, then return to the front and share those thoughts with everyone along with my reflections. Not only does this open the possibilities for congressive people to participate without pressure or fear, it also neutralises ingressive behaviour, because when the potential for performance and a public assertion of superiority is transformed into a human interaction then somehow the ingressive behaviour dissipates. As it happens, it also means I interact with far more women than in a normal 'ingressive' question time, in which I hardly ever see any women ask questions. Sometimes the non-men in the audience just want to tell me about a part of the talk that they really appreciated. This gets filtered out in normal question times, especially when the chair sternly reminds the audience that 'questions end with a question mark'. This well-meaning attempt to preclude ingressive ranting typically just means that the ingressive ranters rant away and then end by saying, '… and my question is: have you thought about that?' and the congressive expressions of appreciation are lost.

Congressive workplace

I realise that in the wider world of work some ingressive structures might be more necessary than in academia, but we can still dream. Promotion might be required because people gradually take on more and more responsibility. One way to make it more congressive might be to make it more about sponsorship and mentorship than about putting yourself forward. My career was substantially and crucially facilitated by various people supporting me along the way, rather than by me putting myself forward – at least, that's how it felt to me. (Incidentally, all the people who put me forward in those career progressions were men.) We do have to make sure this doesn't just sound like old boys' networks. In *Inclusive Leadership*, authors Charlotte Sweeney and Fleur Bothwick warn of the perils of creating support networks that too quickly become exclusive in turn, as can happen – alas – whenever excluded people get together and emulate their excluders by excluding others in turn.

It is also important that this should not be about some sort of arbitrary 'sisterhood'. Women's groups can be off-putting to some women and also seem exclusionary of men. They also too often neglect intersectionality and just end up creating a new hierarchy of power inside them, with straight, middle-class, able-bodied, cisgendered white women at the top imposing their power on all the other women. Instead we could make congressive groups for congressive people and people interested in congressivity.

Another aspect of a congressive workplace would be flexibility, so that everyone can work to their own strengths rather than to an imposed structure. This is a way to get much more out of everyone, as we've seen in the case of Dame Stephanie Shirley, and yet in my institutional academic job allowing people to 'play to their strengths' was seen as unfair.

In *Becoming*, Michelle Obama writes of working to remedy the imbalance in her law firm which hired people who were predominantly male and white. She had to persuade the recruiting team to look beyond the usual metrics of prestigious universities and exam results, and consider their background to understand whether they'd coasted on privilege or raised themselves up from difficult beginnings. I would now call that thinking congressively rather than ingressively about hiring.

Different industries have very different characters as driven by their particular fields, but another one that I know quite well, other than maths and education, is music.

Music

Music and creative arts are fundamentally congressive as they are about creating things rather than about right-and-wrong situations. When musicians get together to make music in a group they are not trying to beat each other. The goal is to make wonderful music together, and that is achieved by everyone contributing and supporting each other.

Unfortunately music is another part of the world that has become rather an ingressive industry, with competitions, stressful auditions, the pressure of performing on stage, ingressive judgement of critics, prizes, and superstar idolisation. In a music competition you contrive scarcity by declaring that only one person can win. Whenever there is a move away from this in which everyone gets a prize, there is an ingressive backlash from people who think that the 'everyone gets a trophy' culture is ridiculous and doesn't breed excellence. The idea that competition produces better results is ingressive, and people who do actually achieve more in competitive situations are more ingressive. The stress of a competition means that you have to have a certain kind of emotional disposition to withstand it, and, as Alfie Kohn points out in *No Contest*, that might be the exact opposite of the disposition needed to play music sensitively.[3]

By contrast, the idea that a supportive and collaborative atmosphere produces better results is congressive, and congressive people may well achieve more in that type of atmosphere. I have founded a consciously congressive music organisation in Chicago called the Liederstube, focused on classical song. There is no barrier to entry – anyone can come and sing, and anyone can come and listen. There is also no barrier between the performers and the audience – everyone sits together in a cosy room with an atmosphere that is more like a party than a concert. It is less like performing (which is ingressive) and more like sharing music we love (which is

---

3 *No Contest* quotes critic Will Crutchfield on this subject.

congressive). There is no emphasis on perfection or correctness (which is ingressive). The pressure of performance on a stage is completely removed and this can result in more emotionally touching music. But it's also more efficient, because when preparing for a performance a huge proportion of the work goes into withstanding stressful conditions.

So for a tiny fraction of the effort of preparing a formal concert performance we can make music that is almost as good except that maybe we play more wrong notes. It's rather ingressive to care about wrong notes anyway; it is more congressive to care about the feelings we communicate and share with everyone in the room. I like the congressive version and think it draws some people into the classical music world who are put off by its ingressivity (and the ticket prices).

I would like to run 'auditions' like this, where it's less like a competition and more like getting to know each other in a genuine, congressive music-making environment. We still find out what we need to know about each other as musicians. It takes a bit longer, but it filters congressively for something that should be congressive.

I invite and urge you to think about any other aspects of life, how ingressive they are, how ingressive they really need to be, and in what ways they could actually benefit from being more congressive. And then I urge you to dream up ways to nudge them in the direction of more congressive behaviour. This doesn't have to be in daunting large structures such as entire industries – it can start with the structures of our personal interactions, such as in any discussions we have with other people.

Congressive discussions

Discussions are a small form of social structure, the smallest being a dialogue between just two people, or even with yourself. Too many discussions are really arguments aiming to establish superiority or dominance. Congressive discussions instead establish a connection between people and greater understanding. Perhaps women tend to do more of the latter and so have more meaningful friendships; of course, referring to gender in that way is likely to be divisive, so it's better to say that congressive behaviour fosters more meaningful friendships and everyone can learn it. (I've given some specific tips on this in the Appendix.)

Unfortunately, along with our general obsession with competitions and 'winning' we are very attached to the idea of winning arguments. In fact, when I do interviews about my previous book *The Art of Logic*, the headline that often goes up involves something about 'abstract mathematics helps us win arguments', despite the fact that one of my major points in the book is that I think it is more productive to seek to understand each other rather than to 'win'.

We see copious examples of ingressive arguments, including formal debates in debate clubs, during elections, and in political slots on news programmes. When politicians are interviewed on air the situation is usually one of mutual attack, but I often find that in that process I don't discover anything more about what the politician thinks or why they think it. Moreover, the politician is rarely genuinely challenged about their position as I think they would be if they were asked to expand on it and talk about it in depth rather than rapidly parry sharp blows from the interviewer.

In non-political interviews there is often a more

congressive approach, not based fundamentally in attack. If a scientist has made a breakthrough worthy of news then the interviewer's aim is more obviously to get the scientist to explain it to everyone (although journalists who are not science specialists do sometimes have a sceptical 'what on earth is the point of this' spin on more esoteric aspects of science).

What if we had congressive political interviews as well? Perhaps we'd learn more and also have more real challenges to our politicians instead of the usual situation where they can just say the same soundbites over and over again or deflect questions by answering something else entirely.

We could learn more from congressive discussions in our daily lives too, instead of having divisive ingressive arguments that don't help us make progress. For example, the so-called oppression olympics take place any time people in a discussion are competing with each other to show that they are the 'most oppressed'. Too often in the environment of identity politics it seems that only the most oppressed person in the room is allowed to have experiences that are considered valid, because everyone else counts as 'privileged'. This is sometimes taken as an argument showing that identity politics is unhelpful, but I think it only shows that *ingressive* identity politics is unhelpful. If we acknowledge everyone's privilege and everyone's misfortune and everyone's relationship to everyone else, then we can work congressively to reduce *all* oppression instead of being distracted by the oppression olympics.

During ingressive arguments I have rarely seen anyone persuaded of anything. We can retrain ourselves to make arguments more congressive by seeking to understand things rather than defeat other people. In those situations you might suddenly find that you don't mind discovering that you're wrong after all, and that you learn a lot more about something. I have learnt about congressive arguments from the most congressive person I know, my friend Gregory, with whom discussions are always about discovery and not right and wrong. I always learn from those discussions and never feel bad; rather, I feel that progress has been made. It might be particularly hard between a woman and more ingressive man, because of the gendered assumptions that still persist in society about who is supposed to play the role of an 'expert'. It still frequently occurs that my very existence as a woman mathematician makes a man feel so insecure that he immediately starts trying to belittle me. Sometimes for my psychological well-being I don't bother engaging, but at other times I can concentrate on taking a congressive approach and then have a more productive discussion in which I seek to understand where his insecurity is coming from, and in doing so he understands more as well, once the ingressive distraction is removed.

And remember that ingressive distraction just benefits the people in power, currently dominated by men, and contributes to the oppression of women and others. Competition obstructs collaboration, and ingressive energy obstructs us from working congressively to change the power structures. It might seem that there's little an individual can do under those circumstances, but here is a very inspiring story of one woman who effected powerful congressive change, starting small.

Restorative vs punitive justice

Becoming Ms Burton is an extraordinary and important memoir of an extraordinary woman, Susan Burton, and her journey to becoming a social activist, rescuing women from a systemically racist criminal justice system in which she was herself trapped for years.

The first part of her story is tragedy upon tragedy. She was abused and raped as a child and teenager, and spiralled into worse and worse circumstances like so many black people in the US who are neglected and exploited on all sides. Her five-year-old son was killed in a hit and run by a policeman and in her grief she fell prey to the crack cocaine wars that were just beginning to rage in LA. There began fourteen years of her bouncing in and out of prison for drug offences, victim of a justice system that sought only to punish women like her, and not to help them.

Eventually she managed to free herself with the help of a residential rehab programme and turned her attention to helping others, first on an individual scale as a carer, and then by setting up A New Way of Life, a charity providing residences for women coming out of prison to help them help themselves. She began small, with just one house for ten women in 1998, and gradually expanded so that to date they have helped more than a thousand women, including reuniting more than 300 of them with their children, according to their website.[4]

She writes compellingly of the injustice of a system that simply incarcerates people who need help. It is even senseless from a financial point of view – she points out that

4 http://anewwayoflife.org/what-we-do/

taxpayers pay up to $60,000 per year to incarcerate a person, comparable to tuition fees at an elite university. This sparked the slogan 'Yale not jail!' – A New Way of Life apparently provides its services at a third of the cost of incarceration. Incarceration typically leads to more incarceration, as people get stuck in a loop that becomes increasingly hard to escape. She writes: 'Jail had done nothing to stop my addiction. Education, hard work, dedication, a support system, and knowing there were opportunities for me and that my life had value: these were what had made all the difference.' What worked was a congressive restorative approach rather than an ingressive punitive one.

An unnuanced point of view is that if you don't commit crimes then you don't get stuck in this loop in the first place. However, Ms Burton's book sets out quite starkly the two-tier system in which white people and black people are treated differently for the same minor offences. Moreover, 85% of women in prison are victims of physical or sexual abuse, and these women are 'disproportionately black and poor'. Many became drug addicts in desperate circumstances, manipulated and exploited by the dealers, but they are treated only as perpetrators, not victims.

A more congressive point of view says that if we helped people more in the first place it would be a much more effective way to reduce crime, and that punishment doesn't reduce these issues as much as help and rehabilitation do. This might seem like a naively utopian dream for the future, but congressive social structures have existed for a long time, just not in white capitalist societies.

Native American cultures

In *The Sacred Hoop*, Native American writer Paula Gunn Allen writes about the social structures in tribes, social structures that were deliberately undermined by white colonial invaders. She describes tribal societies as egalitarian, organising events and phenomena not by hierarchies as in white (Christian) societies, but with everything related to everything else in a harmonious or unified way. She even writes of traditional war that it was 'not practiced as a matter of conquest or opposition to enemies', but as a ritual to gain the attention of supernatural powers.

She describes Paul Le Jeune leading the Jesuits to try and convert the Montagnais to Christianity, 'How could they understand tyranny and respect it unless they wielded it upon each other and experienced it at each other's hands?' Le Jeune was determined to change the minds of these people, who 'did not punish children, encouraged women in independence and decision making, and had a horror of authority imposed from without', and who did not believe in people assuming superiority over other people: congressive values. She describes the development of tribes over 500 years of colonisation as they have progressively moved from gynocentric egalitarian social systems towards the European, patriarchal, hierarchical style. I would say: from congressive to ingressive.

Allen writes: 'Under patriarchy men are given power only if they use it in ways that are congruent with the authoritarian, punitive model.' I would extend this to say that in an ingressive society people are given power only if they use it in ingressive ways. As we have discussed, this has encouraged everyone of all genders to be as ingressive as possible in our ingressive society, or alternatively it has caused those who don't want to be ingressive to give up and resign themselves to being powerless, or latch onto some borrowed ingressive power from someone else (for example, by marrying a powerful man). But Allen shows that it was not like that before the arrival of the white colonisers. It appears that congressive society isn't something new we need to build, but something old we need to restore.

Congressive democracy

Democracy is a fundamental way in which we interact with the ingressive power structures of society that the European colonisers spread around the world. Any form of democracy is of course more congressive than a dictatorship, but different nuances of ingressivity and congressivity occur within the different types of democracy that exist around the world, both in how the democratic systems operate and in the election processes used to select politicians. I think our systems of power are still predominantly ingressive. Of course, politics around the world is still largely dominated by men, and so this is another area in which a more congressive approach could change that status quo. What might that mean?

Considering systems of democracy, direct election of a president (as in the US) is much more ingressive, whereas in other systems such as in the UK the head of the government is just the leader of whichever party has the most seats. It is possible that this is why it is taking much longer to get a woman president of the US than a woman prime minister of the UK or chancellor of Germany.[5] But I would

---

5 See, for example, Zack Beauchamp, 'The US has a female

prefer to say that really it makes it harder to get a congressive leader.

Whatever the system, election campaigns are increasingly ingressive. It's true that there is genuine competition involved as there is a limit to the number of people we can elect, thus the scarcity of candidates is real. However, there are more and less congressive ways of going about this selection process, as can be seen by looking across different systems in the world and different times in history.

They are perhaps more congressive in countries other than the US, for example in the UK, where spending is limited by law and negative campaigning is less prevalent (although that seems to be changing for the worse everywhere). Election debates have become extremely ingressive. There is a video still available of a US Primary debate in 1980 between Republican candidates George H. W. Bush and Ronald Reagan, in which they essentially agree about wanting a compassionate approach to immigration from Mexico. This sort of polite, compassionate consensus seems unthinkable now, especially on that topic.[6]

A two-party system is another ingressive aspect, as is the First Past the Post (FPTP) voting system, also called 'Winner Takes All', which makes it sound even more ingressive. In fact, FPTP voting reduces the impact of small parties, so tends to cause a *de facto* two-party system. It's ingressive because it's basically like a zero-sum game – to win more

---

presidential nominee for the first time. Here's why it took so long', *Vox*, 26 July 2016.

6 See, for example, https://www.youtube.com/watch?v=YsmgPp_nlok.

votes you have to make someone else lose them (unless you can rely on motivating substantial numbers of previously apathetic people to vote).

Voting systems are a tested example of how a more congressive structure can affect people's behaviour and influence them in a different direction, which is just the sort of thing I believe we should implement more widely.

More congressive voting systems than Winner Takes All exist, such as proportional representation, or systems with multiple rounds or a ranked choice. In these, votes for smaller parties aren't so obviously wasted, as they can still count towards a proportion of seats, or, in a system with multiple rounds, candidates who are knocked out will have their votes redistributed according to the rankings of the voters. One such system that is already in use in some parts of the world is Alternative Vote, also known as Single Transferable Vote, Ranked Choice Voting, or Instant Runoff. The idea is that if there are many candidates then the one with the fewest first choice votes is eliminated, as in a multi-round knockout challenge, and everyone votes again. To avoid the logistical complication of everyone *actually* voting again, voters are allowed to rank all the candidates on their ballot in their order of choice, so that if their top choice is knocked out their vote can then be redistributed to their next choice. This process of knocking people out and redistributing votes continues until someone has a majority.

Some people object to ranked voting systems on ingressive grounds, claiming that it means the 'winner doesn't necessarily win'. However, this is only true if you define the winner to be the person with the most first-choice votes. So all that the ingressive person is objecting to when they say 'the winner doesn't necessarily win' is that winner-takes-all and ranked voting might produce different results – which is the whole point.

It is hard to see how to move to any such congressive systems when the current system gives disproportionate power to certain parties. Those parties hold power and benefit from the status quo, so it is particularly hard to know how to change that except by slowly influencing more people to be congressive and to believe in congressive structures. Some places in the US have managed to vote in this system, including the state of Maine, and several cities such as San Francisco, Minneapolis and, most recently at time of writing, New York City. The case of Maine is quite interesting: Ranked Choice seems to have been successfully voted in as part of a reaction to Governor Paul LePage being elected with much less than 50% of the vote twice in a row. It seems he has called himself 'Donald Trump before Donald Trump became popular',[7] and the more moderate vote was split between a Democrat and an Independent. Indeed, apparently the more a Maine voter supports Trump, the less they support Ranked Choice Voting.[8]

But supporting a congressive system shouldn't just be about wanting a particular person to win; it's not about manipulating the outcome to be the one you want, but about believing that a more congressive process is more representative. That said, if ingressive people will only support the system for self-serving reasons, it might help once they see

---

7 Katharine Q. Seelye, 'Paul LePage was saying whatever he wanted before that was a thing', *New York Times*, 13 August 2018.
8 Joseph Anthony, 'Ranked-choice voting is a partisan affair in Maine', *Bangor Daily News*, 24 September 2019.

that their own vote could be split, not just their opponents' votes. The UK voted against such electoral reform in the referendum of 2011 (by a substantial margin), but the political climate of the 2019 election meant that the votes on the left and the right both seemed in danger of being split – on the right between the Conservatives and the Brexit Party, and on the left between Labour and the Liberal Democrats. However, in the event, the Conservatives benefited from vote-splitting without suffering much from it, so the chance for all sides to support a more congressive system for self-serving reasons once again faded.

Aside from voting systems and party systems, the whole adversarial style of modern politics might be regarded as an impediment to sorting anything out. Persuading people to be less adversarial in a system that rewards it is likely to be futile, and so the key would be to change the system into one that doesn't reward such behaviour. Again, it's a way we could make a congressive structure to nurture, rather than force, congressive behaviour.

This is another potential benefit of ranked voting systems: they encourage a less combative and partisan campaign because candidates need to appeal to their opponents' supporters in order to pick up second-choice votes. Such systems also increase the chance of having several smaller parties who need to cooperate with each other in order to get anything done. One more congressive system involves 'Constitutional conventions, cross-party forums and citizens' assemblies' to supplement parliament.[9] I would call

---

9 John Coakley, 'Brexit has nearly broken British politics. Here's how to fix it', *Guardian*, 30 March 2019.

this a more congressive approach to running a country. Of course, the question remains of how we change the status quo to achieve that.

Changing the status quo

People naturally resist changing the status quo if the old system has played out particularly well for them, as with political parties who benefit from the Winner Takes All voting system. When people have enjoyed disproportionate power, they will resist having it taken away.

But, curiously, I have not yet met much outright objection to the notion of ingressive and congressive behaviour from men; rather, men have shown a lot of enthusiastic support and sometimes relief in response to these ideas, which to me is testimony to one of the benefits of ungendering our argument. Of course, it's possible that I just don't meet many of the men who are really holding disproportionate power. However, I have encountered objections from women who have been very successful by learning to emulate ingressive behaviour. Perhaps they haven't imagined another possibility or are afraid they will be worse off in the new stuctures.

The status quo is evidently supported by those who win in the current structures (of whom there are not that many), but unfortunately also by those who are convinced that they could win, indeed that they might soon win. Perhaps this is an essential aspect of the 'American Dream', the idea (or rather, the dream) that in this structure anyone can rise up, even all the way to the top. Convincing people that they are about to rise up while simultaneously keeping them down is a particularly pernicious way to maintain power.

Unfortunately, most people who think they're about to win are deluded, like addicted gamblers, because most people do not win in an ingressive situation. But the winners benefit from those who continue to delude themselves that they can win. In the case of gambling in casinos, the winners are the casinos themselves. The casinos depend on people being convinced that they can win, but the only way to be sure of winning, unless perhaps you're a world-class poker player, is to get out of the casino and do something else.

I think a new approach based on congressive values will be better for the whole of society, including men, women, and oppressed minorities. It just won't feel better for the people who currently hold undue amounts of power and who want to cling to it. It's also important to remember that there are problems separate from this that we must still address, including problems of prejudice based explicitly on gender, race, and other identities. But the idea is to separate those out from the questions of character that we have been discussing throughout.

Aside from whether it changes gender imbalances, I believe it's the right approach for the good of society, inclusivity and equity. But I do think that making more congressive systems and structures could lead to better outcomes about gender imbalances, with a less divisive dialogue. It's my congressive feminist dream.

'Smash the patriarchy!' is the somewhat ingressive battle-cry of some branches of feminism. I prefer to say something more congressive and less divisive: let's transform our world into a more congressively led future. Let's reward, encourage and nurture congressive behaviour and build congressive structures. At the same time, let us stop fabricating ingressive stuctures like unnecessary competitions and contrived zero-sum games, and stop using ingressive filters for congressive roles.

I think that this is an implicitly feminist aim. The ingressive structures and our ingressive attitudes are perpetuating power structures that favour men, so anything we do to dismantle those structures is, I believe, an act of feminism. And just as bias can be explicitly against women specifically because of gender, or implicitly through unconscious associations or structural issues, I think dismantling it can also be explicit or implicit.

Explicitly dismantling gender bias means changing those structures along the one-dimensional gendered lines. This is still an important part of dismantling explicit sexism, but will not address indirect and structural bias in the system.

I believe that the new dimension of ingressive and congressive traits can help us overcome bias that can't be addressed on the dimension of gender alone. I think this is how we can deal with implicit bias in the system that comes from our association of character with gender, and indirect bias that comes from favouring ingressive behaviour; but it will also have an effect on explicit bias, as it is a way for everyone to escape a one-dimensional gendered mindset and think more clearly about what contributions to society we want to make. With every individual who escapes that thinking, the hold of both implicit and explicit gender bias will be lessened.

It is hard for congressive people to change the balance of power while everything favours ingressive people, but we can do it congressively by collaborating and supporting each other. This is different from the arbitrary 'sisterhood' that some feminists think all women should be in, which is in itself a divisive idea. Many women rankled at the idea that they 'should' vote for Hillary Clinton 'because she is a woman'. The divisive idea of doing something for someone 'because she is a woman' can now become doing something for someone because they are congressive. And I think that helping congressive people in certain ways is not just fair but also has the possibility of huge benefits for everyone.

We may be losing all sorts of congressive people from further education, graduate study, research, high-level jobs and promotions by not giving them the external validation they need *and deserve*. Thus we leave all of those opportunities to the ingressive people who base their self-esteem on nothing in particular and so might well be less skilled at the task in question, less willing to improve, and less able to grow.

I hope that we will all get together and think of more and more congressive systems for the good of society. We particularly need action from those in any positions of power and influence, including teachers, especially teachers of subjects that are traditionally rather ingressive such as science and mathematics. And we need more appreciation of traditionally congressive subjects such as the creative arts.

We need action from people running companies and hiring, to find ways to make their companies more congressive if they believe in inclusivity and getting the best out of everyone. We crucially need action from parents raising children so that the next generation need not be quite so ingressive, right from the start. We could also do with some academics to do formal research on this, from the point of view of psychology, sociology, and more.

You might think that my congressive utopia is a lovely dream but can never become reality. Well, I always say that if we declare that something is impossible then that will become a self-fulfilling prophecy. We don't have much to lose by trying, and who benefits if we don't? The people who benefit the most are the ingressive people who are currently in power. As Alfie Kohn says, 'I would prefer to see skepticism directed at the status quo rather than employed in its service.' Jessa Crispin reminds us that 'breaking away from the value system and goals of the dominant culture is always going to be a dramatic, and inconvenient, act'.

We don't have to change the entire world at once, and we don't have to work on all scales at once. We can start small, by making personal changes in our daily lives and interactions with individuals. We can change the way we discuss things, and we can even just stop using implicitly ingressive language such as the word 'competitive' to mean 'good' and slang phrases like 'for the win' as a compliment. We can encourage congressive behaviour in the people around us and practise neutralising ingressive energy wherever we encounter it.

We can then move on to small micocosm utopias like my congressive classroom and my music salon. This could include family units (as Philippa Perry suggests), groups of friends, clubs, small organisations and companies. We could then move on to bigger institutions and companies, whole schools and universities, and gradually on to structures in society like entire industries, election processes and democracies. We started with a small step, a small shift in perspective away from the dimension of gender. We started with a few case studies, a few role models. I believe that we can gradually roll out these ideas to the rest of the world.

Thinking along the dimension of ingressive and congressive traits enables us to clarify our thoughts and solve particular problems, as a good theory should. In fact I feel that thinking this way has given me radical clarity about how to live my life, how to contribute to the world, and how to be a good feminist, working to change power structures while still recognising the experiences of individuals and the range of other dimensions on which they enjoy advantage and suffer disadvantage.

I think that what is at stake here is nothing less than our humanity itself. One of the major features we have as humans is our congressive ability to communicate and collaborate. This is how we have formed communities at the scales we have, and made progress in ways that no other animal has. But there is also a major ingressive aspect of humans that sets us apart: that we enslave people (and other animals), exploit them, and exert power over them.

It's not a competition, but perhaps if there is only one competition it is between the ingressive and congressive forces in humanity. I hope that we will choose congressivity, and work together, congressively, to build a better future for everyone.

Postscript

A week into our coronavirus lockdown in March 2020, I heard a compelling story from my dear friend Amaia. A poet, optimist, and believer in nature, she was looking for hope in the midst of this global crisis, and found something quite unexpected: a fascinating example of social change following crisis, involving a troop of savanna baboons in Kenya, studied by Robert Sapolksy and Lisa Share.[1]

Natalie Angier writes about this research, introducing the idea of positive change coming out of a crisis: 'Freak cyclones helped destroy Kublai Khan's brutal Mongolian Empire, for example, while the Black Death of the fourteenth century capsized the medieval theocracy and gave the Renaissance a chance to shine.'[2] As a more recent case, the Second World War was an atrocity that must never be repeated but during it women showed their worth as fully participating workers in society, and society was forced to acknowledge it and change at least some of its ways.

The baboon story, in short, is that an aggressive and hierarchical baboon society transformed into a nurturing and

---

1 Sapolsky, R. M., Share, L. J., (2004) 'A Pacific Culture among Wild Baboons: Its Emergence and Transmission', *PLoS Biol* 2(4): e106.
2 'No Time for Bullies: Baboons Retool Their Culture', Natalie Angier, *The New York Times*, 13 April, 2004.

peaceful one in the space of one generation. The catalyst for the change was a crisis in which half of the males died of tuberculosis. Crucially, it was the most aggressive alpha males who died: there had been a fight with a neighbouring tribe over some food at a garbage dump. Only the most dominant males had been able to hold their own in the fight and get to the food, but the food turned out to be contaminated with tuberculosis.

The surviving society consisted of previously subordinate males, the females and the young. Sapolksy and Share describe the transformed society, and I would call it congressive. It remained congressive even when the males who had survived the epidemic had died and new males from outside the troop had joined; the new males learned the congressive behaviour from the existing members. The researchers found that although the society was still hierarchical, it was less aggressively so, with the dominant males less prone to taking out their aggression on the subordinate males and the females, and with more 'affiliative' behaviour such as mutual grooming to build community. And this new congressive community was better for everyone in terms of their levels of stress, without compromising their ability to survive as a troop.

I am absolutely not suggesting that we hope to wipe out all the ingressive people in the world with an epidemic. But, perhaps, like the baboons we can respond to a crisis with a cultural shift away from our hierarchical, gendered, ingressive culture, towards an inclusive, congressive future.

This global pandemic of 2020 might show us some ways in which traditionally celebrated ingressive behaviour really is ineffective at best, and deadly at worst. Risk-taking and disobeying social distancing instructions increase the spread of infection. Basketball player Rudy Gobert publically scoffed at the risks, touched all the microphones at a press conference, and then became the first NBA player to test positive, resulting in the abrupt shut-down of the season. Politicians accustomed to bold posturing have found that you might be able to out-posture a human enemy but you can't out-posture a virus. As industries have closed down, one by one, and a large proportion of people's income streams has been cut off, even the right-wing conservatives have acknowledged that people need help from the government, possibly even a universal basic income – albeit it temporarily – in order to survive. Suddenly congressive policy that was dismissed as utopian even on the Left two weeks ago is being proposed by the Right as a matter of urgency.

At more individual levels, those of us who are confined to our homes are finding more ways to share: ideas, skills, expertise, our art. Arts institutions have been opening up their digital archives, streaming free concerts, making virtual tours. Some people have been hoarding toilet paper, but others have been hand-sewing face masks for hospitals desperately short of supplies and overwhelmed by the influx of patients.

Perhaps by the time you are reading this it will all be over; I certainly hope so. I hope that by the time you are reading this we are no longer confined to our individual homes, that we have gone back to eating and drinking together, going to concerts, visiting galleries, browsing in real physical bookstores. But I also hope that we have not forgotten this slight tilt towards congressivity that I am starting to see. Rather, I hope that we have built on it, and realised that these ideas generalise beyond a crisis, and that even when times are not so universally dire, we can think and act congressively to make life better.

Appendix

How to be more congressive

Here are some practical suggestions for how to become more congressive in your daily life. We could think of it as congressive personal training for anyone who wants to unlearn ingressive behaviour and learn to be more congressive. Most of this is an exercise in gradually shifting emphasis rather than (ingressively) suddenly flipping your entire behaviour. Remember that the congressive approach is to keep practising rather than aim for a particular outcome, and acknowledge that it's an ongoing process rather than something that can be suddenly achieved. If it doesn't quite work on one occasion it doesn't mean you failed; the only way to fail is not to try. If you at least try, then whatever happens you will have more case studies to think about for the future.

Look for similarities between people and situations rather than differences

This is like the idea of seeking to empathise with other people rather than showing how different you are. This might be in response to someone talking about their experiences, their tastes, their dreams, or things that are bothering them. Remember that it's not a competition. If someone talks about something bad that happened to them, it's not very congressive to 'sympathise' by saying something similar happened to you, but worse. Showing sympathy when something very bad has happened is tricky, because if you say that you understand, it might be invalidating to someone who feels that what they are experiencing is uniquely terrible. So seeking to validate someone's experiences is a very important aspect of congressive behaviour, and this requires empathy to tell whether they want to know that their experience is unique or want to feel that their experience is universal. For example, with logistical frustrations it can be very calming to know that we are all sharing these frustrations as a community, and it can be very isolating to feel that nobody else is bothered by these things. Not to mention, if everyone is bothered by the same things we're more likely to have a chance of changing them.

Seek to support others rather than advise them, unless they specifically ask for advice

Unsolicited advice is ingressive as it assumes a position of superiority – you are taking it upon yourself to assert that you know better than the other person does about what is going on, and you are assuming that there is something you can offer that they won't already have thought of. It also assumes that they are looking for a solution, and that a solution exists, which has the danger of invalidating the realness of their problem. Perhaps what they're really looking for is sympathy and acknowledgement that the problem is indeed annoying, frustrating or upsetting. If you give them advice under those circumstances they are quite likely to resist it and tell you why they think it won't work, or say they've tried it, and escalate their expression of how bad they feel. It's like in Aesop's fable of the sun and the wind, when the wind blowing harder just makes the person pull their cloak on tighter, whereas the sun shining brighter induces them to take the cloak off.

You might feel that you are a problem solver and you really want to solve that person's problem for them. One way to try and convince yourself (or an ingressive friend) to support rather than advise is to observe that the described problem isn't the actual problem that needs fixing – the 'problem' is the need for validation. So if you really want to help fix the problem, providing sympathy and validation is in that sense solving the problem.

If you really think you might have something helpful to offer, a congressive way to do so would be to say something like, 'Would you like to hear something that helped me?' and then be ready to stop if they say no. In the end a congressive approach is to help someone solve their own problem instead of swooping in and thinking you can fix it for them.

Build mutually beneficial situations

There are too many zero-sum situations in the world, often unnecessarily, as with the contrived scarcity we discussed that causes unnecessary competitions. Even if we can't control large structural situations we can all make changes in interpersonal situations.

Alfie Kohn discusses more abstract forms of scarcity as well, including interpersonal situations of 'mutually exclusive goal attainment', that is, in which each party involved can only achieve their goal by preventing the others from achieving theirs. That is an ingressive situation. This often happens in manipulative or abusive relationships, where the abuser deliberately creates a zero-sum game, convincing the other person that the only way to make their partner happy is to be hurt themselves.

To be more congressive we can build situations for mutually beneficial goal attainment instead. Even ordinary conversations have potential for this as they often turn into zero-sum games unnecessarily, with a limited resource of 'being right'. If a conversation is instead about gaining understanding, then, as with education, the resource is not limited. In fact, more understanding can be gained if everyone collaborates and shares the efforts rather than obstructing everyone else's efforts. Instead of a zero-sum game we have one where the outcome is 'more than the sum of its parts'. It is no longer an ingressive win/lose situation at all, but a congressive mutually beneficial one. Other mutually beneficial situations include playing to strengths in relationships (romantic or professional, or just between friends) so that everyone contributes what they're good at.

Better still, we can resist and actively dismantle zero-sum situations that other people create. This might seem ingressive but there are congressive ways to do it. Sometimes I manage it by convincing myself that every time I defuse an ingressive situation I'm helping the world. I also remind myself that the zero-sum situation was not my fault in the first place, so resisting it need not be considered my fault either.

Adopt the principle of charity

This means seeking the most generous interpretation of someone's point rather than an antagonistic simplification. The latter is usually a precursor to a straw person argument. This is part of a broader point of trying to understand what someone really means rather than jumping down their throat and immediately trying to demonstrate that they're wrong. We can always look for the sense in which people are right rather than point out ways in which they're wrong.

For example, if your partner says, 'You never do the washing up!' you might immediately point out that one time three months ago when you did the washing up when they were ill in bed. That would be an ingressive response. A congressive response would be to observe that they probably don't mean that you literally never do the washing up, but rather that they're frustrated by how little you do the washing up. You might then think about whether it's worth doing more washing up to help them feel better.

This is a good congressive approach whenever somebody makes a point that riles you. This can happen on both sides of a disagreement about bias and oppression. If someone expresses frustration about 'men' or 'white people', try to understand that they don't necessarily mean all men or all white people, but think about the structures built into society and consider whether you can see any truth in the fact that men and white people hold a lot of power. This is relevant in any discussion with someone more disadvantaged than you in some way. On the other hand, if you are the more disadvantaged person expressing frustration, and the man or white person (or white man) gets cross with you for being 'prejudiced right back', you could congressively try to think about why they are feeling that way. It might just be coming from a selfish fear that if everyone were treated fairly then they themselves would not, in fact, be given so much power. But, if you have the energy, see if you can determine whether their frustration is coming from their own feeling of exclusion from power in society on the grounds of social status, wealth, sexuality or something else.

Congressive role play

Here are some real examples of ingressive things that people have said to me, and three types of response I have thought of as I described with a diagram at the end of Chapter 6. This is something you can always try doing after an interaction has occurred if you find yourself regretting how it went. The more you practise thinking up congressive responses slowly by yourself, the easier it will become to slow down and do that in person as well.

'*Typical woman!*'

Ingressive	Typical man!
Passive	[Silence]
Congressive	That was hurtful.

'*It's sweet that you believe in logic.*'

Ingressive	Too bad you haven't grasped the concept of what logic actually is.
Passive	Well I guess I'm an optimist.
Congressive	What do you mean by that?

How to be more congressive

'*Women are just vain.*'

Ingressive	Well men are just idiots.
	You could stand to look in the mirror once in a while.
Passive	No we're not.
Congressive	That was rather belittling.

'*You're not very scientific are you.*'

Ingressive	Said the pot to the kettle.
	So when was your last scientific paper published?
Passive	I'm being perfectly scientific.
Congressive	Which particular part of my argument do you disagree with?

'*Congratulations! You earn so little you might as well quit.*'

Ingressive	Right, and just be a career failure like you.
Passive	But it's better than nothing.
Congressive	Are you trying to be supportive?

'*One day you'll learn.*'

Ingressive	One day you'll retire.
Passive	Maybe.
Congressive	What do you think I'll learn?

'*That is not how we do things.*'

Ingressive	Well no wonder this place is a disaster.
Passive	Oh I'm sorry.

Congressive   I'm interested to know what your reasons are.

'Oh yeah? What makes you think you're successful?'

Ingressive   Well I don't have an insecurity complex like you seem to.

Passive   I think I'm successful in my own way.

Congressive   Do you think there's only one form of success?

'Spoken like a true Chinese.'

Ingressive   Spoken like a true racist.

Passive   Huh?

Congressive   What exactly do you mean by that?

Acknowledgements

This book has been an even bigger and more daunting journey than the previous one, and I was crucially helped by innumerable people along the way, by their insight, wisdom, experience, support and inspiration.

First I would like to thank Andrew Franklin and all at Profile Books for their deeply congressive ongoing support. Heartfelt thanks are also due to Lara Heimert, TJ Kelleher, and all at Basic Books. I am immensely grateful to have publishers on both sides of the Atlantic who continue to believe in me as I push my boundaries far beyond where I originally saw them. For this book I must again particularly thank my wonderful editor Nick Sheerin, whose vision and empathy combine to help me say what I'm trying to say in a way that will reach further, wider and deeper.

I owe many thanks to my students at the School of the Art Institute of Chicago. Their intellectual energy and challenging curiosity directly led to this book. I would also like to thank everyone at the School. It is a revelation to work for a supportive, congressive institution.

None of this would happen without the support and inspiration of my parents, my sister, and my little nephews

Liam and Jack, who are rather less little than when I thanked them in my previous book.

Thanks are also due to my wonderful friends who have provided my congressive support network for my whole life, and in particular during this phase of my life in which random insults from around the world are unfortunately quite likely. I could not do all this without that support. Special thanks are due to those whose thoughts and experiences I have specifically referred to in this book: John Baez, Edray Goins, David Kung, Marissa Loving, Ruth Jurgensen, Emily Riehl, Sandmeyer's, and the three hundred or so singers who have thus far sung at the Liederstube. Thanks to Ben Hamill for undertaking some congressive personal training with me.

Thank you to my personal cheerleaders Amaia Gabantxo and Jason Grunebaum, and my brilliant assistant Sarah Ponder. As always I am grateful to Sarah Gabriel for being a beacon to cut through my brain fog.

I would also like to thank all those, too many to name, who have given me an opportunity to speak on this subject at various universities around the world in the last few years while I was developing these ideas. The response from audience members helped me to clarify my thoughts and push them further.

I am grateful for my congressive school education, which celebrated contribution to the community more than 'high achievement', although I admit I was a bit bitter about that at the time; I now appreciate the values I learnt. I am also grateful to all those who helped my career move forward when I might otherwise not have believed in myself enough: my directors of studies Paul Glendinning and Jan Saxl,

Acknowledgements

my PhD supervisor Martin Hyland, and also Paul-André
Mélliès, André Hirschowitz and Peter May.

Last and most, this wouldn't have happened without
Gregory Peebles, who has to be known to be believed, and
even then is figerally incredible.

Index

Index

Index